MW01411045

Praise for *The First Green Wave*

Canada's environmental movement has a rich and significant history but has very few historians who have taken the time to chronicle and preserve that history. In this lively account, Ryan O'Connor has contributed enormously toward remedying that gap. Pollution Probe was one of the very first of Canada's environmental groups. Those early activists charted a course that many still follow – and more should."

 – Elizabeth May, OC, Leader of the Green Party of Canada

O'Connor's work provides crucial insights into the origins of one of the key organizations in the modern environmental movement. It is a must-read for any student of environmental policy and politics in Canada.

 – Mark S. Winfield is a professor in the Faculty of Environmental Studies at York University and the author of *Blue-Green Province: The Environment and the Political Economy of Ontario*

Ryan O'Connor has not only captured the facts regarding the early modern environmental movement in Canada, but the spirit of those days as well. For me that spirit will always be reflected in the image of my best friend and founder of Pollution Probe, Tony Barrett, wearing an army helmet and headset, fiddling with a military switching device he installed on his phone, and shouting "Incoming! Incoming!" I have often been asked whether I'm not a little embarrassed about those early days. My answer: "Never. We were right about absolutely every issue we tackled and have since been proven so." A breezy read, for seminal times.

 – Monte Hummel, OC, President Emeritus, World Wildlife Fund Canada

The First Green Wave is a deeply researched, fine grained, nuanced history of Canada's early environmental movement. Ryan O'Connor convincingly demonstrates that organizations such as Toronto's Pollution Probe pioneered a distinct form of regionally focused, politically centrist environmentalism that reflected Canada's political, geographic, and social realities. It is a fine contribution to the growing body of literature on this important topic.

> – FRANK ZELKO is a professor in the Department of History at the University of Vermont and the author of *Make It a Green Peace! The Rise of Countercultural Environmentalism*

Based on a wealth of research, *The First Green Wave* provides a valuable contribution to the history of Canada's environmental movement. O'Connor deftly examines – often through the voices of key participants – the birth and changing fortunes of Pollution Probe. This is a fascinating account of a social movement, borne of the cultural ferment of the 1960s, and sustained by the pragmatic cooperation of student and professional activists, scientific experts, the media, and business interests. Read O'Connor's book: "Do it"!

> – JOHN F.M. CLARK is the director of the Institute for Environmental History and a lecturer in the School of History at the University of St. Andrews

The First Green Wave

The Nature | History | Society series is devoted to the publication of high-quality scholarship in environmental history and allied fields. Its broad compass is signalled by its title: nature because it takes the natural world seriously; history because it aims to foster work that has temporal depth; and society because its essential concern is with the interface between nature and society, broadly conceived. The series is avowedly interdisciplinary and is open to the work of anthropologists, ecologists, historians, geographers, literary scholars, political scientists, sociologists, and others whose interests resonate with its mandate. It offers a timely outlet for lively, innovative, and well-written work on the interaction of people and nature through time in North America.

General Editor: Graeme Wynn, University of British Columbia

A list of titles in the series appears at the end of the book.

NATURE | HISTORY | SOCIETY
GENERAL EDITOR: GRAEME WYNN

The First Green Wave

Pollution Probe and the Origins of Environmental Activism in Ontario

RYAN O'CONNOR

FOREWORD BY GRAEME WYNN

UBC Press · Vancouver · Toronto

© UBC Press 2015

All rights reserved. No part of this publication may be reproduced, stored in a retrieval system, or transmitted, in any form or by any means, without prior written permission of the publisher, or, in Canada, in the case of photocopying or other reprographic copying, a licence from Access Copyright, www.accesscopyright.ca.

23 22 21 20 19 18 17 16 15   5 4 3 2 1

Printed in Canada on FSC-certified ancient-forest-free paper (100% post-consumer recycled) that is processed chlorine- and acid-free.

Library and Archives Canada Cataloguing in Publication

O'Connor, Ryan, author
    The first green wave : pollution probe and the origins of environmental activism in Ontario / Ryan O'Connor.

(Nature, history, society)
Includes bibliographical references and index.
Issued in print and electronic formats.
ISBN 978-0-7748-2808-6 (bound). – ISBN 978-0-7748-2810-9 (pdf). – ISBN 978-0-7748-2811-6 (epub)

    1. Environmentalism – Ontario – History – 20th century. 2. Pollution prevention – Ontario – History – 20th century. 3. Ontario – Environmental conditions – History – 20th century. I. Title. II. Series: Nature, history, society

GE199.C3O26 2015   333.720971309'045   C2014-905839-X
                                                  C2014-905840-3

Canadä

UBC Press gratefully acknowledges the financial support for our publishing program of the Government of Canada (through the Canada Book Fund), the Canada Council for the Arts, and the British Columbia Arts Council.

Cover illustration: This photograph was taken in Toronto's High Park in conjunction with the launch of Pollution Probe's namesake publication, which was released by New Press in 1970. In tree (left to right): Jack Passmore, Stanley Zlotkin, Paul Tomlinson, Rob Mills, Varda Burstyn, unidentified child, Terry Alden (standing in tree), Monte Hummel, Tony Barrett, Peter Middleton, Brian Kelly, James Bacque (publisher, New Press), and Christopher Plowright. Standing in front of the tree is Donald Chant. *Courtesy of Merle Chant.*

UBC Press
The University of British Columbia
2029 West Mall
Vancouver, BC V6T 1Z2
www.ubcpress.ca

In memory of Siobhan Lily O'Connor

Contents

List of Illustrations / xi

Foreword: Yogi Berra, William Wordsworth, and Canadian Environmentalism / xiii
Graeme Wynn

Acknowledgments / xxiii

Abbreviations / xxv

Introduction / 1

1 *The Air of Death* and the Origins of Toronto's Environmental Activist Community / 12

2 The Emergence of Pollution Probe / 39

3 Building an Environmental Community / 75

4 Probe's Peak / 103

5 The Changing Environmental Landscape / 122

6 Beyond the First Wave / 153

Afterword / 174

Notes / 178

Bibliography / 208

Index / 219

List of Illustrations

1 Filmmaker Larry Gosnell / 15
2 Canadian Broadcasting Corporation newsman Stanley Burke interviewing dairy farmer Ted Boorsma / 20
3 Members of Group Action to Stop Pollution promoting its first meeting, December 1967 / 32
4 Public inquiry into the apparent pesticide-induced death of ducks off of the Toronto Islands, July 1969 / 46
5 "How would you like a glass of Don River water?" advertisement in the Toronto *Telegram*, 29 September 1969 / 53
6 The Don River funeral, held on 16 November 1969 / 54
7 Simon Greed, a wealthy industrialist portrayed by Pollution Probe member Tony Barrett / 54
8 Placement of a wreath at the conclusion of the Don River funeral / 54
9 Pollution Probe meeting with Robert Stanfield, leader of the Progressive Conservative Party of Canada, 1970 / 61
10 Brian and Ruth Kelly, members of Pollution Probe's Summer Project '70, who travelled throughout cottage country / 67
11 Pollution Probe buried a time capsule outside the present-day John P. Robarts Research Library, October 1970 / 79
12 Pollution Probe's telephone directory recycling drive in Metro Toronto, April 1971 / 86

13 Pollution Probe speaker Bob James at a pollution conference, June 1970 / 105
14 Protest against the construction of the Darlington Nuclear Generating Station, June 1979 / 141
15 Staff at Energy Probe, following their split from the Pollution Probe Foundation / 146
16 The Is Five Foundation's curbside recycling programs in Toronto, the late 1970s / 148
17 Pollution Probe's new home, Ecology House, opened in 1980 / 160
18 Total Recycling began an experimental recycling project in Kitchener, September 1981 / 165
19 Total Recycling employee Bob LeDrew stacking blue boxes in Kitchener, 1983 / 166

FOREWORD

Yogi Berra, William Wordsworth, and Canadian Environmentalism

by Graeme Wynn

According to baseball great Yogi Berra, watching teammates Mickey Mantle and Roger Maris hit back-to-back home runs, game after game, for the New York Yankees in the early 1960s was like déjà vu all over again. The phrase – one of many "Yogiisms" that led *The Economist* to recognize Berra, in 2005, as the "Wisest Fool of the Past 50 Years" – springs to mind as I read Ryan O'Connor's *The First Green Wave* – not because this story has been told before – it has not – but because so much of this book deals with times, places, and events that I recall.[1]

Pollution was in the air when I began my academic career in 1968. Late that year, as I settled in to my MA program at the University of Toronto, the release of the Hall Commission report on *The Air of Death*, a television program first aired on CBC in 1967, re-animated discussion about the accuracy of the documentary's claims. As debate raged over Hall's critical findings, Larry Gosnell and Stanley Burke, the producer and narrator of the controversial film, showed their work on the university campus. Once again, viewers saw images of black smoke belching from an industrial plant and heard the grave, familiar voice of CBC television's national news anchorman intone: "Every day your lungs inhale fifteen thousand quarts of air and poison." Continuing, Burke drove home the frightening message: "You breathe sulphur dioxide, which erodes stone. Benzopyrene makes cancer. Carbon monoxide impairs the mind ... Death has been gathering in the air of every Canadian city. Poisons continue to accumulate and you must keep breathing."[2]

Within days, Sherry Brydson, the editor of the University of Toronto student newspaper, reflected on the implications of the Gosnell-Burke documentary in three articles in the *Varsity* – the first of which asked: "Pollution: Is There a Future for Our Generation?"[3] Within weeks, an organization called Pollution Probe began to make its presence evident on campus. Because their offices (first in the cramped quarters of the *Varsity* and then in the Ramsey Wright building) were only a stone's throw from Sidney Smith Hall where Geography was located, their existence and activities were impossible for me to miss. Later in 1969, mere weeks after I completed requirements for the MA, Pollution Probe garnered national attention by staging a funeral for the Don River, the unusual event with which O'Connor begins his book.

A few years later, while in the Maritime provinces pursuing my doctoral research, I encountered one of the most redoubtable of Canadian heroes in the ongoing war against pollution – the everyday, everywhere terror that threatened our way of life. This was Captain Enviro, summoned from the fertile imaginations of Robin Edmiston (writer) and Owen McCarron (illustrator and owner of Halifax-based Comic Book World) at the behest of the environment ministers of the three provinces. According to the garishly illustrated, eponymously named comic book that recounted Enviro's achievements, King Sewage III and his people on the planet Polluto, located millions of miles from earth, faced a crisis. They were endangered by the spread of cleanliness through their territory. Anxious, King Sewage dispatched three trusty lieutenants to find a new home for the Pollutians. Earth seemed to offer decent prospects, so Sludge, Smog, and Slime descended on the disgustingly clean and beautiful Maritime Provinces to take them for their own. There, however, they met our superhero, who looked like a humble sanitation worker but possessed extraordinary powers that enabled him to survive the effects of a magic mind pollutant administered by the Pollutians, stop the polluting rampage of the filthy fiends from afar, and remind everyone of the need for constant effort to keep the region "Land, Air and Water Pollution-free."[4]

As I recall, and O'Connor demonstrates, only a small fraction of the University of Toronto's 25,000 students were sufficiently moved by Brydson's words to write the *Varsity* or attend the meetings from which Pollution Probe emerged. But many more realized that there was something amiss with the air we breathed and perhaps the water we drank and began to ponder the nature of things to come. Others were angered that the university chancellor was a director of the company that operated the fertilizer plant that *The Air of Death* had indicted for poisoning the environment around Dunnville, Ontario. These were heady days in which to be a student. Controversies, protests, and proto-revolutions vied for attention on the Toronto campus as they did on many others, and whether it was spawned by American involvement in Vietnam, the power of the military-industrial complex in the United States, a conviction that mass consumption had run out of control, or more specific concerns about the manipulation of authority and the pervasiveness of pollution, there was widespread concern about the state of the world. But there was also a strong sense of liberation, of the democratization of power, and of new possibilities as students challenged university authorities to implement more democratic governance and more open curricula, and society to change course. Or as the English Romantic poet William Wordsworth had it in "The French Revolution as It Appeared to Enthusiasts at Its Commencement": "Bliss was it in that dawn to be alive, / But to be young was very heaven!"[5]

Yet, I come to *The First Green Wave* with a sense of trepidation. There is something disconcerting for the historical scholar about revisiting the scenes of one's youth through others' depictions of them. Memories and interpretations seem to differ, meanings to shift, and old verities to crumble. Well I remember the astonishment I felt some years ago, when I discovered that an oft-cited expert source of information in a well-regarded doctoral dissertation on the contamination of the Great Lakes was one of my graduate student contemporaries. Him? An authority? How could this be? We make stories of our lives. Others, fellow travellers on parts of our journeys, have their own stories about those points where paths and memories intersect, shaped as ours are, by events before and after, the search for significance and coherence, and perhaps a desire to present a certain face to the world.[6] Then there are historians, magpie-like collectors of shiny objects from the past, who choose and display those objects in their own stories about times/places/events because they fit, or hold together, a particular view of things. "Oh! Pleasant exercise of hope and joy!" (Wordsworth again).[7] How then should I engage O'Connor's interpretation of a thin slice from my own past, his interpretation of a burnished fragment from a rich and oft-recalled period? Perhaps simply by recognizing that memories are fickle, that histories are the creations of their authors (rather than mirrors of a fixed reality) and that Yogi Berra knew what he meant, or meant what he knew, when he said: "You can observe a lot just by watching."[8]

Ryan O'Connor, a young scholar whose environmental sensibilities were forged on Prince Edward Island in the 1990s – when acid rain and ozone depletion, major international issues, were the focus of concern – comes to the subject of this book neither as participant in, nor observer of, the events of which he writes, but as an historian. This gives him a certain distance – greater than mine – from his subject, and he has written what is, in many ways, a very traditional history tracing the rise, and changing fortunes, of a single organization through some two decades. He has fulfilled the historian's scholarly obligation to search out and examine a wide range of primary documents, and he has interviewed key participants in the events he describes. From these sources, O'Connor has constructed an accessible narrative that traces the story of Pollution Probe's emergence and evolution in an engaging manner.

This is an important contribution in its own right. Remarkably, given the considerable volume of writing about Greenpeace, the almost contemporary ENGO that emerged in Vancouver, Toronto's earliest environmental organizations have received little attention from scholars, and perhaps in consequence, general histories of the 1960s in Canada place far more emphasis on the social than the environmental upheavals of that tumultuous long decade.[9] Thus *The First Green Wave* offers a new and interesting story, well told and nicely amplified in the endnotes, many of which add intriguing facets to the central text. Simply put, this book is a significant addition to the literature on Canada in the second half of the twentieth century.

The main lines of O'Connor's account can be summarized thus. Pollution Probe emerged from the commotion generated by *The Air of Death* and subsequent efforts to question the accuracy of that documentary. It was a grass-roots, student-led movement that won important support from a cluster of similarly concerned university faculty members. Those involved in the organization in the early days spanned the political spectrum, but most of the group's key decision-makers were "pragmatic centrists." They endeavoured to establish good working relationships with the business community and politicians, even as the organization developed its reputation as an environmental stakeholder representing the interests of the environment and "the public" of Ontario.

Active, energetic, committed, and skilful in drawing support, Pollution Probe's founders quickly achieved both visibility and success.[10] By the end of 1970, the group had over 1,500 members and had mounted effective campaigns against the pollution of the Great Lakes by phosphates, the air by a 700-foot high "superstack" proposed for a local power-generating plant, highway verges by litter, and cottage country by sewage and gasoline exhaust. It had also published a 200-page guide to living an environmentally friendly life. Pollution Probe affiliates began to proliferate, across Ontario, east into the Maritimes, and west onto the Prairies and were joined by offshoot organizations such as the Canadian Association on the Human Environment and the Canadian Environmental Law Association. By 1972, Probe had established teams to organize and address their activities and concerns across a range of fronts, which included education, energy and resources, recycling, and land use. There was also an Action Team, to organize high-profile events; and a Caravan Team, to take Pollution Probe's message across the province.

An economic recession in 1973 brought a chill to all aspects of the environmental movement as politicians, business leaders, and the public alike made the reinvigoration of economic growth a major priority. Buffeted by circumstances to the point of financial disarray, Pollution Probe was forced to adapt – and shrink. A focus on toxic waste issues and the safety of the drinking water supply brought the organization out of the doldrums in the early 1980s. That decade brought success – even if somewhat indirectly in the development of Pollution Probe's long-standing arguments for recycling into a very successful blue box program by Resource Integration Systems and Laidlaw Waste Management Systems – and controversy – over support for Loblaw's new biodegradable disposable diapers – to the organization. But the landscape of environmental activism was changing away from the local issues that were the main focus of its early protests to the transnational problems that drew attention – and spawned different, pan-Canadian activist organizations – in the last decade or two of the century. Pollution Probe survived into the new millennium and continues, yet with a much narrower focus – and far less name recognition – than in days of yore.

These are the bald "facts" elaborated on, contextualized, and interpreted in the pages that follow. They are in some simple sense the "truth" about Pollution Probe's history, the bare bones fleshed out in O'Connor's telling of a story that one brief account presents as a tale of "an exuberant, attention-grabbing, loosely structured collective of young environmental activists" who believed that a little "humour never hurts," who married guerrilla theatre with science "to raise public awareness and push for change," and whose legacy is an organization working "quietly behind the scenes."

But it is wise to pause here, to ask what histories such as *The First Green Wave* really do. Surely it is more than grubbing up relevant pieces of information and ordering them chronologically. Several paragraphs above, I suggested that O'Connor's youthfulness endowed him with a certain perspective on the subject of his book, a "distance" greater than mine because I was caught up in the stream of time from which Pollution Probe emerged. This was a deliberate nod to a commonplace of historical scholarship that "truth is the daughter of time." I was invoking the notion of "historical distance": the idea – as the great Marxist historian Eric Hobsbawm once expressed it – that "retrospectiveness is the secret weapon of the historian," because it allows him or her to see events in proper proportion and to assess their implications accurately.[11] As an undergraduate, I was instructed in the implied corollary: that it was impossible to study history, properly construed, of the period subsequent to one's birth (some sticklers drew the line well before that) – and that to consider anything of more recent vintage was to dabble in sociology, politics, or some other such amusement. But the last fifty years have taught us many things, and few would hold to this ancient verity today. Rather, we are inclined to say, as has Canadian intellectual historian Mark Salber Phillips, that "what histories 'most manifestly' do is to mediate the relationship between the now of the present and the then of both past and future."[12]

Seen in this light, *The First Green Wave* is a timely and valuable product of its time. It looks back to one of the pivotal moments in the development of Canadian environmentalism when Pollution Probe emerged from a groundswell of concern about the future of life on earth and exercised a great deal of influence, with other important environmental non-governmental organizations, on the development of environmental policy. It does so from a now-point in which there is growing fear (among environmentalists at least) that their influence on policy is waning. Despite the "Inconvenient Truths" marshalled and promulgated by Al Gore, despite glacial retreat, escalating rates of ice-sheet melt, rising CO_2 concentrations in the atmosphere, and repeated warnings that anthropogenic climate change is a real and present danger, rates of fossil-fuel consumption continue to rise, and many governments are unwilling to sacrifice (as they typically see it) economy for ecology.[13] All of this begs big questions about the future of the planet, prompts introspection about the current tactics of environmentalists, and leads one to wonder whether the past can provide the present with lessons that might help secure the future.

Does *The First Green Wave* offer anything that might help us to address these pressing concerns? First it offers hope. Like all good historical scholarship, this book reveals that history is contingent – which is to say that outcomes depend on particular prior circumstances (which are in turn the result of antecedent conditions, and so on, ad infinitum). Had something, anything, anywhere along the line been different, this or that particular historical outcome might not have been what it was. Had Sherry Brydson spent reading week 1969 in Florida rather than Toronto, there might have been no Pollution Probe and the 5 million or so Ontario households that participate in the bluebox recycling program today might be generating three times the garbage they currently contribute to the province's waste stream. The world is, and always has been, a wonderfully interconnected, interdependent place. Contingency makes it highly unlikely that history will repeat itself, but acknowledging its importance encourages us to think deeply about the past, present and future, and their interconnectedness. Acknowledging contingency also creates room to believe that things can be different and reminds us that even small individual actions can have large effects in shaping human (and indeed environmental) affairs.

One of the central motifs of O'Connor's story is that Pollution Probe struggled to retain its relevance and influence as the world and the nature of its most pressing environmental problems changed. The organization began with a strong commitment to local action. The dirty Don River, the belching Robert Hearn superstack, befouled recreational lakes, litter-strewn highways – these were all issues with evident causes and tangible consequences that ordinary people might take action to resist or prevent. Pollution Probe found its cause and made its name addressing what students of pollution call end-of-pipe issues. They fit like a glove with the great environmental mantra of the 1960s and 1970s: "Think globally, act locally." In the 1980s and 1990s, however, the hot-button issues were acid rain – which seemed to come from the clouds (shifting nebulous affairs that drifted in from afar) and fall everywhere – and the so-called ozone hole(s) – far away, undetectable to the human eye, and reportedly caused by the release of invisible molecules (CFCs). These threats, O'Connor suggests, posed a different order of problem, one with which Pollution Probe was not well attuned to deal and that spawned new environmental organizations given to different forms of campaigning.

Thinking about these developments as part of the continuum between past and future may provide helpful perspective on responses to perhaps the most pressing of current environmental concerns: global climate change. Like the hot-button issues of the 1990s, this is a problem that is at once everywhere and nowhere, real yet invisible. The scientific evidence of anthropogenic climate change is unimpeachable, but tangible everyday experience offers no convincing confirmation: bitter winter weather, harsher than any in recent memory, can be accounted for in ways consistent with the atmospheric models of climate scientists but the abstract notion of global warming fails to cohere with the experience of those consigned to shovelling snow late into the spring. Yet, the Montreal Protocol of 1987, a decisive step forward, addressing the ozone hole issue by controlling the use of CFCs, held while the Kyoto Protocol has proven ineffective in limiting atmospheric emissions. Do we need a new approach to campaigning on climate change? Is there a scale question to be addressed here? The success of the Montreal Protocol turned in large part on the availability and relatively low cost of replacements for CFCs. Is the current, generally tardy, response to warnings about impending climate (and ecosystem) tipping points a reflection of the expense and inutility of replacements for the gasoline-powered automobile? For fossil fuel generated electricity? For heavy dependence on global trade? Is the problem in the pocketbook of individuals, or is it a tragedy of the commons scenario: Why deprive oneself of convenience and comfort while others (people/nations) who refuse to do so reap benefits, and collapse or catastrophic change is unavoidable in the (not-so-distant) end? Are we looking at policy failure, corporate greed, the triumph of particular ideologies? Ultimately, the answer is, most probably, a combination of all of these things and more.

How then to move forward? Ryan O'Connor's history gives no direct answer, but for those interested in the Canadian environmental movement and the future of Canadian and global environments, one of the most intriguing and important claims of *The First Green Wave* is its contention that Pollution Probe influenced environmental policy development in Ontario from the 1970s by inserting itself into the existing bipartite model of policy formulation. On this account, environmental policy was traditionally developed in "private negotiations between regulators and firms," between the government and business.[14] Beginning as early as 1971, representatives from Pollution Probe were invited to serve on municipal and provincial policy committees as recognized stakeholders on behalf of the environment. This was possible, argues O'Connor, because the pragmatic centrists in Pollution Probe sought from the outset to establish a working, rather than adversarial, relationship with bureaucrats and businessmen, and this stance ultimately won the group the respect that made many of its accomplishments possible.

This is no small thing, especially in comparison with the history of environmental activism on Canada's west coast, where confrontation between business, government, and environmentalists was the norm through the 1990s. Not for nothing were the protests at Clayoquot Sound in 1993 dubbed the "War in the Woods." Disruptive actions, civil disobedience, staged arrests, and carefully prepared news-bites were the stock-in-trade of environmental activists ("protestors") in British Columbia and many other parts of the country from the 1960s. Not until the final stages of the Clayoquot confrontation was this approach nudged aside by a more collaborative, respectful form of engagement and bargaining involving government, corporate, indigenous, and environmental interests.

Reaching this point was not easy, as the parties involved had demonized one another consistently. It was also, at least in one telling, substantially contingent on the chance meeting of two forceful antagonists from the corporate and environmentalist sides of the struggle in a Vancouver neighbourhood, where they recognized each other as mothers with babies, rather than as sworn enemies; as mothers they had shared interests and thus found common ground. Meeting later for coffee, "mom to mom, person to person," in the words of prominent activist Tzeporah Berman, "they saw each other as people for the first time," and the result was "a breakthrough for forest politics in Canada."[15] To make a complex story simple, this shift in consciousness not only helped to resolve immediate tensions between arch-enemies, it laid the foundations for the Great Bear Rainforest Agreement, reached early in the new millennium by consultation and consensus among First Nations, government, environmental groups, and industry leaders and widely touted as breaking new ground in efforts to tackle "the highly complex and critical problems societies are increasingly coming to face around the globe."[16] If there is a lesson in all of this, and the pages that follow, it may be that hasty judgments, dismissive labelling, and a failure to listen all get in the way of solutions – or that inclusive, open, and respectful discussions in which evidence trumps ideology and conciliation is valued over confrontation, offer the best prospect of realizing a just and lasting future for both humans and nature.[17]

Those who remember Pollution Probe's crowded makeshift offices at 91 St. George Street might find a certain poignancy in the arc traced by O'Connor's history of the organization. Recalling a movement spawned by a combination of youthful enthusiasm and noble idealism (albeit tinged with a certain anxiety) that quickly achieved great things, they might well lament how times have changed. To read *The First Green Wave* is to learn, after all, of a once enormously influential, widely recognized, and youthfully exuberant organization pummelled by changing circumstances that has slipped into relative invisibility in the landscape of early twenty-first century Canadian environmentalism, its place, and this new scene, symbolized by its current headquarters location in a suburban Toronto office tower.

Yet, this is too forlorn a view. Pollution Probe survives, and despite scant knowledge of its origins and achievements among current staffers, and the relative narrowness of its contemporary agenda, it remains politically non-partisan and committed to building partnerships among stakeholders to address important environmental concerns. This is not failure. Nor, though, is it necessarily the future. Environmental problems, socio-economic and political circumstances, and the nature of environmental activism have all been transformed over the last half-century, and there is every prospect that they will be again. With Ryan O'Connor's full and valuable account of Pollution Probe's past in hand, we are surely better equipped to look back as we go forward, not in anger, nor with any sense of nostalgia, but certainly with a firmer understanding of the contingent quality of human affairs, and the hope that flows therefrom: that it is possible for individuals, and groups of like-minded souls, to change the world.

Looking back on the events of 1789 in France, Wordsworth marvelled at how "the inert / Were roused" and both "the meek and lofty" found "helpers to their heart's desire," as they were summoned to "exercise their skill,"

> Not in Utopia, subterranean fields,
> Or some secreted island, Heaven knows where!
> But in the very world, which is the world
> Of all of us, – the place where in the end
> We find our happiness, or not at all!

Pollution Probe may have changed and blended into a landscape in which there is rather less wild-eyed environmental activism than once there was, but as that wisest of fools once said: "It ain't over 'til it's over."[18]

Acknowledgments

THIS BOOK IS THE PRODUCT of years of research, writing, and rewriting. My deepest appreciation is extended to Alan MacEachern, who provided guidance from the get-go. He saw the potential when I presented the idea of working on "green politics" in Canada, and provided the necessary support throughout the research and writing stages. Roger Hall likewise provided support throughout the process, and, when necessary, helped open doors to private collections and interviews. They, along with Finis Dunaway, Jason Gilliland, and George Warecki, read earlier drafts of this manuscript and provided many helpful comments and suggestions for improvement. I would also like to thank Stephen Bocking, Claire Campbell, Peter Campbell, Ian Dowbiggin, Edward MacDonald, Robert MacDougall, Bill Turkel, and Robert Wardhaugh. You all helped with the evolution of this project and my development as an historian. Thank you.

Staff at the University of Western Ontario's D.B. Weldon Library, the Archives of Ontario, Library and Archives Canada, and the University of Toronto Archives helped me find the research materials I needed. My appreciation is also extended to Michael Moir and the staff at the Clara Thomas Archives and Special Collections at York University, David Sharron at the Brock University Special Collections and Archives, and Jessica Blackwell at the University of Waterloo Special Collections and Archives for their help securing images for this book.

This book draws extensively on oral interviews that I conducted with members of Ontario's environmental community and their contemporaries. I wish to extend a sincere thank you to each person who shared their memories with me. Of special note, I would like to thank Merle Chant and Peter Middleton, who shared my enthusiasm for this subject and helped me track down veterans of the environmental movement in three different countries. In addition, I was granted special access to private papers by Merle Chant and Denise Gosnell, while Patty Chilton and Bob Oliver provided access to Pollution Probe's internal archives.

Financial support from the University of Western Ontario, a Social Science and Humanities Research Council of Canada Doctoral Fellowship, and several Ontario Graduate Scholarships, made the research behind this book possible. I would like to thank the following, whose generous contributions helped defray the costs of publication: Sherry Brydson, David Crombie, David Estrin, Monte Hummel, the Love family, Peter Middleton, Robert Mills, Terry O'Malley, Ann Rounthwaite, Mark Rudolph, Lynn Spink, and David Wood, as well as the Pollution Probe Foundation for its administrative support. Additional financial support was provided by the Symons Trust Fund for Canadian Studies.

A version of chapter one first appeared as "An Ecological Call to Arms: *The Air of Death* and the Origins of Environmental Activism in Ontario" in *Ontario History* 105:1 (2013). I would like to thank the Ontario Historical Society for granting permission to republish the article here. I appreciate the comments I received while preparing this article, as well as the comments I received on chapter drafts and conference presentations.

Working with UBC Press has been a positive experience. My appreciation goes out to Randy Schmidt, Megan Brand, and Graeme Wynn. Thank you for your support and professionalism throughout this process. I would also like to extend my gratitude to the anonymous reviewers, whose comments have made for a better book.

Research trips were made possible by friends and family who opened their homes to me, including Mike O'Connor and Amanda Firth, Jenia O'Connor and Mark Moore, and Ryan MacDonald. Friends within the University of Western Ontario history department shared their valuable insight, provided advice, and made life more interesting. At the risk of exclusion, I would like to thank Sandra Ceccomancini and Adrian Gamble, Adrian Ciani, Bill Couchie, Mark Eaton, Devon Elliott, Michelle Hamilton, Rollen Lee, Jeremy Marks, Forrest Pass, Andrew Ross, and Carly Simpson. My deepest gratitude is extended to my girlfriend, Kate Cruikshank, who has provided support and a measure of calm throughout the publication process. Thank you for your patience. Finally, I would like to thank my parents, George and Linda O'Connor, for their ongoing support, and for instilling in me a love of learning and an appreciation for the natural world.

Abbreviations

ACE	Advisory Committee on Energy
CAHE	Canadian Association on the Human Environment
CBC	Canadian Broadcasting Corporation
CCAR	Canadian Coalition on Acid Rain
CELA	Canadian Environmental Law Association
CELRF	Canadian Environmental Law Research Foundation
CFCs	chlorofluorocarbons
COPE	Council Organized to Protect the Environment
CRTC	Canadian Radio-Television Commission
EMR	Department of Energy, Mines and Resources
ENGO	environmental nongovernmental organization
ERCO	Electric Reduction Company
GASP	Group Action to Stop Pollution
IFF	Is Five Foundation
LIP	Local Initiatives Program
MTARC	Metropolitan Toronto Airport Review Committee
NFB	National Film Board
OAPEC	Organization of Arab Petroleum Exporting Countries
OWRC	Ontario Water Resources Commission
RCO	Recycling Council of Ontario
RIS	Resource Integration Systems
TZPG	Toronto Zero Population Growth
UCC	Upper Canada College
WWFC	World Wildlife Fund Canada

The First Green Wave

Introduction

It was an unusual sight. On Sunday, 16 November 1969, the members of Pollution Probe, an upstart group of environmental activists, gathered to orchestrate a funeral for the Don River. Beginning with a solemn hundred-car procession from the University of Toronto's St. George campus, the mourners disembarked at the Prince Edward Viaduct. Standing on the banks of the Don, a crowd of two hundred watched as a coffin – representing the waterway – was carried out of a makeshift hearse. A forty-minute ceremony was conducted by a university chaplain. While expressive grievers, some of them clad in period costume, wailed, descriptions of the river's former grandeur were read aloud. As the event came to a conclusion, a Pollution Probe member portraying the Dickensian industrialist Simon Greed, decked out in top hat and tailcoat, began to extoll the virtues of unhindered economic progress. In a moment of poetic justice, Greed was pied in the face, a wreath was placed in the river, and the mourners dispersed.

Theatrical events such as this have become routine within the environmental movement. International organizations such as Greenpeace and Friends of the Earth have full-time staff dedicated to dreaming up publicity stunts sufficiently novel to capture the interest of an increasingly indifferent public. By contrast, Pollution Probe's event had little trouble garnering attention. Occurring at a time before environmental protests became commonplace, the mock funeral attracted the attention of media outlets throughout the city and across Canada.

My environmental consciousness developed in the early 1990s. As a youth, I was well acquainted with the dangers of ozone depletion, acid rain, habitat destruction, and resource exhaustion. Recycling programs were available in most Canadian cities, environmental studies programs were offered at numerous universities, and environmental journalism had become firmly established. Environmental issues had elbowed their way onto the political agenda, and, although it would not enjoy any measure of political success for many years, the seeds of the Green Party of Canada had been sown. Just twenty-five years earlier, none of these features were present. As Monte Hummel, a former Pollution Probe employee and the current President Emeritus of World Wildlife Fund Canada (WWFC), has characterized the domestic scene, "in 1969 there were no ministers of the environment, no environmental protection acts, and pollution was a brand new word."[1]

Canada's environmental activist community developed rapidly. Only a handful of organizations existed in 1969, but by 1971 they operated in all major, and most minor, cities across the country.[2] These activists have played an important role in raising and addressing issues concerning air and water quality, the use of diminishing resources, toxic substances, and energy alternatives. Nonetheless, very little has been written on the origins and development of environmental activism in Canada. Doug Macdonald, G. Bruce Doern, and Thomas Conway have examined the governmental response to Canadians' environmental concerns. Jennifer Read, Arn Keeling, and Mark J. McLaughlin demonstrate the inculcation of environmental values among Canadians during the 1950s, 1960s, and early 1970s. Philip Van Huizen's work highlights the role of cross-border dam projects in shaping British Columbians' ecological values, as well as the provincial government's efforts to cultivate an environmentally friendly image. And an entire cottage industry has developed for publications concerning Greenpeace, the internationally renowned environmental activist organization that originated in British Columbia in 1971. As Frank Zelko has established, however, Greenpeace never had a domestic Canadian focus, and its rapid evolution into an international organization suggests that it is hardly an ideal organization to study in order to understand the nature of Canada's environmental movement.[3]

This book examines the origins and development of environmental activism in Toronto, home to Canada's earliest and most dynamic community of environmental nongovernmental organizations (ENGOs). At the heart of the story is Pollution Probe. Founded in February 1969 by students and faculty at the University of Toronto, Pollution Probe quickly established itself as a leading force within the nascent Canadian environmental community.[4] Emphasizing the core ideals of sound science, public engagement, and an effective utilization of the media, as well as the necessity of accessing the corridors of power, it met with success in its first year of operation, which saw well-publicized confrontations with Toronto's City Hall over the reckless use of pesticides, with Ontario Hydro over air pollution, and with the detergent industry over the phosphate-caused pollution of the Great Lakes. These actions, which inspired the emergence of Pollution Probe affiliates across Canada, were just the beginning for the organization. In later years, it addressed a wide range of issues, from waste reduction to its pioneering work on energy policy and inner-city environmental justice, often pushing the boundaries of what was considered a matter of environmental concern. Pollution Probe would also serve as a mentor within the Canadian environmental movement, fostering the growth of additional institutions while also sharing its expertise on effective lobbying and fundraising with other organizations, such as British Columbia's Scientific Pollution and Environmental Control Society (later renamed the Society Promoting Environmental Conservation) and Halifax's Ecology Action Centre.

Pollution Probe provides insight into the early history of the Canadian environmental movement. In the United States, the environmental activist organizations evolved from existing conservation groups. This came as a result, beginning in the 1950s, of a gradual infusion of ecological values and a newfound activist orientation in groups such as the Sierra Club and the Audubon Society. Furthermore, the old-line conservation groups used their large membership bases and financial clout to launch new ENGOs such as the Environmental Defense Fund (1967) and the Sierra Club Legal Defense Fund (1971). As Christopher Bosso explains succinctly, "the founders and patrons of most environmental advocacy organizations [in the United States] were other organizations."[5] This scenario did not repeat itself in Canada. Although Canada was home to numerous conservation organizations, including the Federation of Ontario Naturalists (1931), the British Columbia Sportsmen's Council (1951), the Conservation Council of Ontario (1952), and the Nature Conservancy of Canada (1962), these groups were highly regionalized and lacked the deep pockets of their American cousins. Furthermore, although George Warecki notes that an ecological consciousness had crept into the Ontario conservation movement during the late 1960s, by that point the first Canadian ENGOs had already begun to organize.⁶ Thus, rather than evolving from pre-existing conservation organizations, Canadian ENGOs such as Pollution Probe appeared on the scene almost spontaneously.

The Canadian ENGOs that emerged during the late 1960s and 1970s were regional entities. This further differentiated the Canadian environmental movement from that of the United States, which, despite a multitude of region-specific organizations, was dominated by large national organizations. This difference is in part attributable to the origins of such groups: as noted, ENGOs in the United States had the support of the well-heeled conservation organizations. However, it is also the direct result of the high costs and administrative difficulty of maintaining a truly national presence in a geographically huge yet sparsely populated country that contains many regional and cultural fissures.⁷

This book examines the environmental movement in Toronto from its origins in the late 1960s to its coming of age in the 1980s. As Robert Paehlke has demonstrated, the movement experienced two distinct waves, as well as an intermittent lull, during this timeframe. The first wave, beginning in the late 1960s, focused on local pollution problems as well as energy and resource issues. In addition to the emergence of Canada's initial environmental activist organizations, the first wave was characterized by heightened sensitivities to ecological issues among the general public and government. By the late 1970s, however, interest in environmental issues on the part of both the general public and government declined. Concern for the environment revived in Canada during the mid-1980s. This second wave of the environmental movement was characterized by the emergence of transnational concerns, such as acid rain, global warming, and the depletion of the ozone layer. The second wave also emphasized the preservation of biodiversity and wilderness areas. These issues were addressed by the creation of a series of new, pan-Canadian ENGOs such as the Canadian Coalition on Acid Rain (CCAR), Sierra Club Canada, and the WWFC.⁸

This book maintains a distinction between the conservation movement, which first arose at the turn of the twentieth century, and the environmental movement, which is a product of the postwar rise of ecology. While interrelated, the movements differed in important ways. As John McCormick explains in *Reclaiming Paradise: The Global Environmental Movement*:

> If nature protection had been a moral crusade centered on the nonhuman environment and conservation a utilitarian movement centered on the rational management of natural resources, environmentalism centered on

humanity and its surroundings ... There was [in the environmental movement] a broader conception of the place of man in the biosphere, a more sophisticated understanding of that relationship, and a note of crisis that was greater and broader than it had been in the earlier conservation movement.[9]

This line of reasoning is echoed by Samuel Hays in "A Historical Perspective on Contemporary Environmentalism":

[The] conservation movement was associated with efforts of managerial and technical leaders to use physical resources more efficiently; the environmental movement sought to improve the quality of the air, water, and land as a human environment. Conservation arose out of the production or supply side of the economy, the environment[al movement] out of the consumer or demand side.[10]

To this point, it is worth noting that first-wave Canadian ENGOs such as Pollution Probe expressed little interest in issues concerning wildlife habitat and the forests. Furthermore, the environmentalists interviewed for this book commonly differentiated themselves from the conservationists, characterizing theirs as a new, youthful movement struggling to protect the planet from pollution and other forms of ecological degradation. This distinction diminished over time as ecological values crept into the former conservation organizations and conservation concerns penetrated the ENGOs. The transition was typified by concern over the disappearance of tropical rainforests in the 1980s; likewise, groups such as the WWFC emerged that straddled the line between conservation and environmental activism. Despite the inevitable degree of overlap between the two forces, the distinction between the conservation and environmental movements is maintained in this book.

Although Pollution Probe was central to the emergence and development of the environmental movement in Toronto, it did not act alone. Numerous organizations emerged between the late 1960s and early 1980s. First on the scene was Group Action to Stop Pollution (GASP). Formed more than a year before the emergence of Pollution Probe, GASP was launched with much fanfare. The Canadian Environmental Law Association (CELA) and the Canadian Environmental Law Research Foundation (CELRF) were founded to provide the movement with a legal arm, while Toronto Zero Population Growth (TZPG) addressed the controversial neo-Malthusian worldview that the ever-expanding human population was at the centre of the global environmental crisis. Still other groups emerged to fill specific niches, including the countercultural Is Five Foundation (IFF), which focused its energies on advancing the cause of recycling, and Greenpeace Toronto, which brought the values of direct action environmentalism, first seen on Canada's west coast, to Ontario's capital. Thus, while Pollution Probe receives the lion's share of attention in this book, it would be impossible to tell the story of the movement's development in the city without incorporating the stories of its contemporaries.

The Canadian environmental movement did not begin with the publication of Rachel Carson's *Silent Spring* in 1962 or the celebration of the first Earth Day in 1970. While historians of the United States cite these as key events in the environmental movement's popular emergence, the origins of environmental activism in Toronto can be traced back to the 1967 television documentary *The Air of Death*. Produced by Larry Gosnell and featuring renowned Canadian newsman Stanley Burke, *The Air of Death* drew attention to the myriad effects of air pollution in Canada. Televised on 22 October 1967 by the Canadian Broadcasting Corporation (CBC), it was not the first documentary to tackle this subject, but it was the first to attract a large audience. Critically hailed, it nonetheless drew the ire of industrial interests, which attempted to discredit the filmmakers and their findings. In the ensuing thirty-two months, the filmmakers were subjected to two high-profile investigations, an Ontario-ordered inquiry, and a Canadian Radio-Television Commission (CRTC) hearing. Chapter 1 tells the story of *The Air of Death,* and will demonstrate how it, and the subsequent controversy, was directly responsible for the creation of Toronto's first two environmental activist organizations, GASP and Pollution Probe.

Chapter 2 examines the Toronto environmental community from *The Air of Death* controversy to the summer of 1970. While GASP enjoyed an initial rush of interest among Torontonians, drawing an estimated three hundred to its December 1967 founding meeting, it never again reached such heights. By 1969 it had morphed into the pet project of an ambitious municipal politician. Lacking any measure of broad-based support, it ceased operations in the summer of 1970. Meanwhile, the student-based Pollution Probe, formed in 1969, found itself thriving. This chapter examines the opposing trajectories of these pioneering Canadian environmental activist organizations and argues that an important difference was that Pollution Probe enjoyed the institutional support of the Department of Zoology at the University of Toronto, which provided the group with credibility as well as the infrastructure necessary to operate full-time. The support of the Department of Zoology was not in itself a guarantee of success, however, as demonstrated by the rather ineffectual emergence of TZPG. Rather, Pollution Probe, unlike GASP and TZPG, benefited from the energies of a relatively large and active membership. Pollution Probe also benefited from the presence of a dedicated cadre from elite backgrounds who played a central role in shaping its organizational character. This group's willingness to approach the business community for support made Pollution Probe unique among ENGOs during the 1970s, enabling it to finance its ambitious plans. Subsequently, it made the transition from a volunteer-driven organization to one run by a full-time paid staff of four. This chapter also describes the emergence of Pollution Probe affiliate groups across Canada – a clear and telling example of the organization's national influence.

Chapter 3 examines the period between autumn 1970 and the end of 1971. During this time, Pollution Probe exerted its leadership and laid the foundation of Toronto's modern environmental community. Buoyed by its early successes, Pollution Probe's staff grew from four to sixteen in September 1970. The staff fostered the creation of CELA, CELRF, and other organizations that complemented their work. The organization also expanded its parameters, from an initial focus on pollution issues to more deep-rooted issues concerning energy and resources. Pollution Probe's continued growth created problems for the group, however. Although the paid staff was responsible for day-to-day operations, key decisions were made by consensus at general meetings. As the number of paid staff and volunteers continued to increase, the decision-making process became more difficult. Marathon sessions became the norm, and calls for a restructuring of Pollution Probe arose. Amid increased tensions, the resignation of a long-time staffer, and the threatened resignation of a second, in summer 1971 the organization created an executive director position, with responsibility for providing oversight of Pollution Probe's operation.

Pollution Probe reached its apogee between 1972 and 1974. As Chapter 4 demonstrates, this period saw continued growth within the organization. Reorganized according to a hierarchical team model, Pollution Probe branched out into a variety of areas not previously associated with environmental activism, including environmental justice and city planning. Besides surveying the organization's varied activities during this period, Chapter 4 examines the internal dynamics of Pollution Probe's staff. However, the ENGO's extended period of growth, dating back to its founding, would soon come to a halt. When Canada entered a recession as a result of the 1973 energy crisis, it became increasingly difficult to secure operational funding, and in 1974 the organization had its first brush with contraction.

Chapter 5 tells the story of the Toronto environmental community's changing landscape through the 1970s. While the energy crisis led to a period of austerity at Pollution Probe, the newfound public interest in energy issues resulted in the launch of a semi-independent sister project, Energy Probe. Long the standard bearer of the city's environmental community, Pollution Probe soon found itself eclipsed by Energy Probe. This, combined with the emergence of the new, high-profile Toronto-based organizations Greenpeace Toronto and the IFF, spelled the end of Pollution Probe's local dominance. Despite the enthusiasm generated by Ecology House and Pollution Probe's new energy-efficient demonstration site-cum-headquarters, ongoing financial difficulties led Energy Probe, the more prosperous of the two organizations, to sever its affiliation in 1980.

As noted in Chapter 6, Pollution Probe managed to reverse its fortunes in the early 1980s through a newfound focus on hazardous waste and public health, including involvement in the high-profile Love Canal court proceedings in New York. Pollution Probe found itself ill-fitted for the period, however. Whereas the late 1960s and early 1970s were marked in Canada by the emergence of localized ENGOs, these were joined in the 1980s by pan-Canadian organizations such as Greenpeace Canada, the WWFC, Sierra Club Canada, and the CCAR. With their broad-based support, these groups were better equipped to address the defining issues of the period, particularly acid rain, the depletion of the ozone layer, and the decline in global biodiversity, which tended to be international in scale. Pollution Probe continued operations, as did other localized ENGOs across the country, but never regained the prominence of its formative years.

This book draws on extensive archival research. Pollution Probe's records are found in two locations. The bulk of the early and mid-1970s material, including reports and correspondence, are located at the Archives of Ontario. Additional material was gleaned from Pollution Probe's internal archives. Other sources include the Omond McKillop Solandt fonds at the University of Toronto Archives; the John Swaigen fonds at the Wilfrid Laurier University Archives; the Energy Probe fonds at the Archives of Ontario; and the Tony O'Donohue fonds at the City of Toronto Archives. I was granted unprecedented access to the papers of Dr. Donald Chant, a key figure in the development of Pollution Probe, and filmmaker Larry Gosnell, whose documentary *The Air of Death* inspired Toronto's early environmental activists. This book also makes extensive use of oral history. Sixty-seven interviews were conducted between 18 November 2007 and

27 May 2010. Those interviewed include activists, academics, politicians, members of the business community, and the media. Thus, the book draws a unique and telling portrait of this long-neglected aspect of Canadian history, weaving together the stories of many organizations and individuals.

This book also expands our understanding of postwar movements in Canada. As sociologist William K. Carroll notes, the late 1960s and early 1970s were "the climax of a period of social movement activism in Canada."[11] In the introduction to *Debating Dissent: Canada and the Sixties,* Lara Campbell and Dominique Clément write that the "sixties were an historical moment that fomented a revolution in education, racial divisions, anxieties about national security, consumerism, Aboriginal mobilization, anti-Americanism, the search for national identity, clashes between capital and labour, innovations in public policy, and debates surrounding the family, health, and the environment."[12] This turbulent period has spawned a growing academic literature. Between 2008 and 2012, three edited collections dedicated to Canada in the 1960s were published. While much was said about topics such as student activism, the New Left, and the expansion and protection of civil rights, not a single chapter was dedicated to the history of Canada's nascent environmental movement.[13] Likewise, the environmental movement is conspicuously absent from such wide-ranging monographs as Myrna Kostash's *Long Way from Home: The Story of the Sixties Generation in Canada,* Doug Owram's *Born at the Right Time: A History of the Baby Boom Generation,* and Bryan Palmer's *Canada's 1960s: The Ironies of Identity in a Rebellious Era.*[14] This leaves an obvious gap in the literature, for if we are to understand the 1960s as "a transformative period,"[15] how can we ignore the emergence of the environmental activists who played a vital role in educating the public and lobbying for protective measures?

On the surface, it makes sense that the Toronto environmental activist community first developed in the late 1960s. As with its contemporary social movements, the environmental activists believed that if people worked hard they could build a better – and in this case cleaner and healthier – future. This belief among baby boomers that they could parlay their unprecedented affluence and hard work into a more ideal world was typical of the spirit of the 1960s. Pollution Probe's roots at the University of Toronto further connect it to the climate of change that was part of the zeitgeist of the decade. With a vibrant student movement that gained in strength and radicalism as the decade wore on, the University of Toronto was home to high-profile protests against the United States' ongoing war in Vietnam, a popular series of teach-ins that attracted thousands to discuss controversial topics such as revolution, human overpopulation, and the role of religion in international affairs, and the emergence of the New Left Caucus, a motley group of Maoist-Marxists.[16] In addition, just blocks away from Pollution Probe's university headquarters were Yorkville and Rochdale College, key gathering spots for Canada's counterculture.

Closer analysis reveals, however, that Pollution Probe was quite different from its campus contemporaries. Most student groups from this period were associated with the political left, and existing scholarship highlights the leftist orientation of Canadian social movements during the 1960s. Such characterization does not fit Pollution Probe. Although there were those among its volunteers and staff who aligned themselves with the left – for example, one of the early members was noted feminist, social critic, and author Varda Burstyn – the organization adopted a centrist approach that made it equally at home dealing with the Toronto business community and the Progressive Conservative government of Ontario led by Premier Bill Davis. As explained in Chapter 2, this was the result of the elite backgrounds of the early leaders of the organization.

Pollution Probe's centrist approach and ease around power brokers also distinguishes the organization from its counterparts at Greenpeace. Founded in 1971, Greenpeace was shaped by a number of features unique to its origins in Vancouver, British Columbia. These defining characteristics, historian Frank Zelko notes, include resonant anti-Americanism, which was inflamed by the city's large population of draft dodgers, the New Left, and the counterculture. These characteristics resulted in a natural eschewal of cooperation with business and politicians. Instead, Greenpeace became a more confrontational organization that gained fame for its utilization of direct action techniques, as in its initial campaign to stop US nuclear tests in the Aleutian Islands by navigating a boatload of activists into the detonation zone.[17] Thus, while Pollution Probe and Greenpeace both focused on environmental causes and effectively harnessed the mass media, their tactics differed considerably due to the circumstances of their founding.

This book also provides insight into the evolution of environmental policy making in Canada. It has been noted that environmental policy was traditionally developed through the bipartite bargaining model, which featured, in the words of Doug Macdonald, "private negotiations between regulators [i.e., government] and firms."[18] In this scenario, the government and its respective agencies were considered the rightful representatives of the environment and the public good. Environmentalists were excluded, notes political scientist George Hoberg, because they lacked the necessary "organizational sophistication and political clout"[19] to warrant an invitation to the negotiating table. According to most scholars, it was not until the environmental reawakening in the mid-1980s that ENGOs and other environmental interests were invited to join government and industry in making policy, a development partly attributable to the fact that government had lost the public legitimacy necessary to bargain on behalf of the environment.[20]

In his recent study of the relationship between the environment and the political economy of Ontario, Mark S. Winfield notes that the collapse of the bipartite bargaining model began in the 1970s. Although he credits the environmental movement with only partial success in creating a new norm of multipartite bargaining concerning environmental matters during this decade, he attributes the progress made to the increasing sophistication and professionalization of Canadian ENGOs that occurred during the mid-1970s.[21] This book contends that the decline of bipartite bargaining, particularly in Ontario, was closely related to the emergence of Pollution Probe and its legal spinoff CELA, and can be traced to the early 1970s.

Repeated surveys and opinion polls have demonstrated that ecological values are entrenched among a large percentage of Canadians.[22] Much of this can be attributed to the activists who have worked hard to educate the public and elected officials on environmental issues. Of course, the integration of environmental values among Canadians is not absolute; one need look no further than the Conservative federal government, which formally withdrew Canada from the Kyoto Protocol, cut hundreds of jobs at Environment Canada, and described opponents of the Enbridge Northern Gateway oil pipeline development as, in the words of Natural Resources Minister Joe Oliver, possessors of a "radical ideological agenda."[23] Meanwhile, Canada continues to be affected by a multitude of environmental concerns, including climate change, the disappearance of wildlife and their natural habitat, as well as air and water pollution. In short, there is still much to be done in the ongoing quest to protect Canadians' environment. It is my hope that this book, in addition to answering various historical questions, will provide some insight into how environmentalism can succeed in the future.

1
The Air of Death and the Origins of Toronto's Environmental Activist Community

O N THE EVENING OF Sunday, 22 October 1967, the CBC pre-empted perennial ratings favourite *The Ed Sullivan Show* in order to broadcast a documentary from its Farm and Fisheries Department. Directed by Larry Gosnell and hosted by national news anchor Stanley Burke, *The Air of Death* was an exploration of air pollution's adverse impact on the environment. Heavily promoted by the CBC, it proved to be a ratings hit as well as a critical success. It also drew the ire of industrial interests due to its allegations of human fluorosis poisoning in Dunnville, Ontario. Subsequently, the film and the team behind it were subjected to two high-profile investigations: an Ontario-ordered inquiry, and a CRTC hearing.

The Air of Death was a pivotal event in the development of environmental activism in Toronto. Before the broadcast, the city was devoid of ENGOs; just sixteen months later, it was home to two environmental activist organizations, both of which attributed their founding to the controversial documentary. It was not the first documentary to raise concerns about Canada's environment, or even the first documentary to address fluorosis pollution in Dunnville. However, a combination of the publicity that surrounded the documentary and the subsequent public inquiries transformed *The Air of Death* into a *cause célèbre* that mobilized the public in a manner previously unseen in Canada, giving rise to the first generation of Toronto's environmental activists.

As Christopher Bosso bluntly notes in *Environment, Inc.*, "Origins matter."[1] In order to understand the operation and development of an ENGO, it is necessary to understand what inspired its creation. Much has been written about the origins of environmental activism in the United States, but the subject has rarely been broached in Canada. While there is no denying that the intellectual current of American environmentalism influenced Canadians, as evidenced by the popularity of such works as Rachel Carson's *Silent Spring* north of the border, this did not launch environmental activism in Canada. Rather, it took a high-profile and shocking exposé of homegrown environmental degradation on the national broadcaster, combined with an obvious effort to discredit the filmmakers, to inspire Toronto's first environmental activist organizations, Group Action to Stop Pollution (GASP) and Pollution Probe.

The Birth of *The Air of Death*

In November 1966, the Canadian Council of Resource Ministers sponsored "Pollution and Our Environment," a five-day conference in Montreal. Conceived as a gathering place for Canada's leading minds to identify key environmental issues, the event attracted over six hundred delegates representing government, industry, and the public, in addition to four hundred observers from across Canada and abroad. Attendance at this conference proved to be a pivotal event in the career of Larry Gosnell, the CBC Farm and Fisheries Department's media delegate. Born on the family farm in Orford Township, Ontario, on 18 May 1923, Gosnell studied agronomics at the Ontario Agriculture College in Guelph. While his work as a radio and television producer focused on social and economic issues affecting rural Canada, much of his early work celebrated the benefits derived from scientific advances in agriculture. By the late 1950s, his tone had acquired a critical edge, and the widespread use of chemical sprays by farmers became a point of interest. He addressed this subject in his 1960 National Film Board (NFB) production *Poisons, Pests and People*, which highlighted the dangers posed by insecticides to humans, farm animals, and plants.[2] According to NFB collection analyst Marc St-Pierre, Gosnell's original script, completed in June 1959, "vigorously denounce[d] the spraying of insecticides," arguing that it presented "a danger to plants, animals and humans." Senior management at the NFB informed Gosnell that his script was unacceptable and demanded a rewrite that accentuated the more beneficial aspects of insecticides.[3] Gosnell made some alterations, and the ensuing production aired in half-hour segments on CBC's *Documentary 60* program in February 1960. Film historian D.B. Jones describes the version of *Poisons, Pests and People* that aired as "journalistic and unengaged" and "not particularly interesting as documentary art."[4] Nonetheless, representatives of the forestry and agriculture ministries deemed the film to be overly critical after it was shown at a natural resources conference in October 1961, and the documentary was quietly removed from the NFB's distribution list.[5]

Despite his early work on the ecological consequences of insecticides, the Pollution and Our Environment conference proved to be an eye-opening event for Gosnell, who later explained that "for me the Conference was a revelation on the degree of pollution that had already happened in our country."[6] On returning to Toronto, he began developing the idea of a three-part primetime television series that would explore air, water, and soil pollution. He faced two major impediments – the subject matter was rather gloomy fare for prime time, and the Farm and Fisheries Department had no experience in producing programming for this valuable time slot – but these concerns subsided when Gosnell recruited Stanley Burke, anchor of *The National News,* to participate in the project. One of Canada's most recognized and respected figures, Burke had a notable background in journalism, having served as president of the United Nations Correspondents Association as well as the CBC bureau chief in such locales as Washington and Paris. Described in the contemporary press as "glamorous" and a "dashing figure,"[7] he was attracted by the urgent tone of Gosnell's project. When asked about his decision to invite Burke's participation, Gosnell would downplay Burke's celebrity and highlight his journalistic credentials. Nonetheless, the addition of Burke's star power proved key to getting the project off the ground. On 25 January 1967, Murray Creed, head of the Farms and Fisheries Department, met with Doug Nixon, the CBC's Director of Television (English Network), and the project proposal was given the green light, with the stipulation that the films must be made interesting enough to maintain the interest of a general audience.[8]

Gosnell set about educating himself on the subject and sought out experts on urban air pollution in Ottawa, Montreal, Syracuse, New York City, and Washington, DC, while dispatching research assistants to the heavily industrialized cities of Windsor, Sarnia, Hamilton, and Detroit. Throughout April, the research concentrated on issues pertaining to urban air pollution. Two vital developments occurred in May. First, it was decided that the still-unnamed special would pre-empt the Sunday night ratings hit *The Ed Sullivan Show* in the autumn lineup, thereby ensuring a sizable audience.[9] The project also took a significant twist when Gosnell attended a lecture in New York City on the topic of fluorosis. There he heard the results of a study of Garrison, Montana, where vegetation, crops,

FIGURE 1   Filmmaker Larry Gosnell won many awards throughout his career, including the Canadian Council of Resource Ministers' Award of Excellence for *The Air of Death* in the 1967 Resources Reporting Awards Competition. *Source:* Personal collection, Denise Gosnell.

and cattle had been devastated by effluent from the nearby Rocky Mountain Phosphate plant. In March 1966, local ranchers had received $123,000 in damages after a court found that the plant's fluorine emissions were at fault.[10]

The Garrison presentation drew Gosnell's attention to the situation unfolding in the vicinity of Dunnville, where farmers were complaining of fluorine pollution from the Electric Reduction Company (ERCO) phosphate plant in Port Maitland. This situation was examined in a segment on CBC television's *Country Calendar* on 26 February 1966, as well as in the 19 October 1966 edition of CBC radio's *Matinee*. Although these segments failed to garner much attention beyond their intended agricultural audiences, they did provide a starting point for Gosnell's research on the topic. Particularly useful was the "Air Pollution" segment on *Matinee*, produced by his old friend Rodger Schwass. As Gosnell later acknowledged, Schwass served as a key source, providing background information and contacts.[11]

The first signs of crop damage related to the ERCO plant were reported in 1961, just three years after it began operating, when Port Maitland farmer Joseph Casina and his customers noticed a significant decline in the quality of his produce. Casina suspected that industrial fumes from the nearby plant might be at fault, so he contacted the Department of Agriculture, which in turn notified the Department of Health's Air Pollution Control Bureau. As the problems continued unabated, Casina struck up a dialogue with W.B. Drowley, director of the Air Pollution Control Bureau, and Everett Biggs, the deputy minister of agriculture, in hopes of determining the root cause of the damage. Despite efforts to measure pollution in the area, the government officials refused to blame ERCO's effluent. The problem worsened. In 1963, area cows began exhibiting symptoms of foot rot. In 1964, Biggs wrote Casina confirming that the "crop damage ... appears to be caused by certain industries in the area." By August, numerous cattle had died under mysterious circumstances, and Casina himself had been hospitalized.[12]

In the summer of 1965, urinary and bone analysis conducted at the Ontario Veterinary College confirmed that area cattle had been afflicted with bovine fluorosis; monitors set downwind of the plant during this period likewise revealed high levels of fluoride residues. As evidence continued to mount that fluoride emissions from ERCO were responsible for the cattle and crop damage, negotiations began between the Ontario Federation of Agriculture, representing the local farmers, and ERCO. In September 1965, the parties agreed to select an arbitrator to assess the value of damages. According to the settlement's guidelines, ERCO would cover the costs of damage to crops, ornamental plantings, and livestock, but only for the current year. Furthermore, before disbursing payments, ERCO required farmers to sign a release acknowledging that payment was not an admission of guilt on the part of ERCO, and that the recipient waived the right to further damages through the end of 1965. The vast majority of affected farmers signed the agreement, either because they felt it was the only available avenue for compensation or because they were compelled by immediate financial need. A total of $86,188.94 was awarded to the farmers in 1965; an additional $112,221.74 was secured for damage suffered the following year.[13]

To this point, attention had focused on the impact of fluoride effluent on farmers' crops and livestock. A more sinister possibility arose in June 1967, when Gosnell met Dr. George Waldbott, a Detroit-based allergist. In the following months, the two discussed the situation in Dunnville numerous times over the phone. Gosnell later described Waldbott as "certainly the most knowledgeable medical man we'd spoken to about fluoride";[14] with the support of local farmers, he invited Waldbott to visit Dunnville on 13 September to discuss symptoms with the locals. Waldbott determined that two of the nine farmers he saw were suffering from fluorine intoxication, a potentially fatal affliction.[15]

Although Waldbott, a native of Germany who immigrated to the United States shortly after earning his medical degree in 1921, was a well-regarded allergist who served on the staff of Wayne State University and two local hospitals, he was a controversial figure within the medical establishment. By the 1950s, his research had begun linking water fluoridation with health problems. Although water fluoridation was one of the period's most contentious public issues, as evidenced by the 136 plebiscites and referenda on this matter across Canada from 1960 to 1966, it had been endorsed by expert bodies such as the Canadian Dental Association, the Canadian Medical Association, and the Royal Commission on Health Services. Waldbott's reports on the dangers of water fluoridation were published in numerous peer-reviewed journals in Europe, but his research was rejected by the major scholarly publications in North America, which led him to develop his own anti-fluoridation infrastructure, including the American Society for Fluoride Research and the bimonthly *National Fluoridation News*. In 1960, he argued for a losing cause before the Morden Commission, which had been established to reconsider the freeze on new municipal water fluoridation programs in Ontario.[16] Gosnell later acknowledged that he knew Waldbott to be an outspoken opponent of water fluoridation, but that this "was a subject in which I had no professional interest."[17] Despite Gosnell's efforts to keep the issues of water fluoridation and fluorosis separate, Waldbott's participation in the making of *The Air of Death* further inflamed an already controversial project.

Gosnell attempted unsuccessfully to arrange an interview with Dr. Roy Pennington, vice president of ERCO's Agricultural Chemicals Division, who admitted in the 1966 *Matinee* "Air Pollution" segment that the farmers' hardships were "at least in part from our operations down there."[18] In the ensuing telephone conversations, Pennington informed Gosnell that he had not received the necessary clearance from his superiors.[19] An 18 March 1969 memo by Dr. Omond Solandt, vice chair of the board at ERCO, reveals that the company feared being singled out in the documentary. As Solandt explained: "I felt that it was very unwise for a small company such as ERCO, which is a very minor factor in air pollution on a national basis, to appear on such a program. Responsibility for representing industry on such a program should be taken by the big industries for whom waste disposal is a major continuing problem."[20]

THE AIR OF DEATH BROADCAST

The Air of Death opened with a stark image of black smoke pouring out of an industrial plant, then cut to video of an expanding human lung, over which Stanley Burke announced in his distinctive drawl that "every day your lungs inhale fifteen thousand quarts of air and poison."[21] As the camera alternated between an old man being tested for a pulmonary condition, a large smokestack, children playing outside an industrial factory, and a hospitalized man with a breathing apparatus inserted through his trachea, Burke continued to set the tone with his voice-over:

> You're an old man in a box or a child at play. You can't choose not to breathe. You must breathe fifteen thousand quarts a day, air and poison. You've got to breathe. You breathe sulphur dioxide, which erodes stone. Benzopyrene makes cancer. Carbon monoxide impairs the mind. They cut a hole in your throat. Death has been gathering in the air of every Canadian city. Poisons continue to accumulate and you must keep breathing.

Burke then appeared on camera. Against the backdrop of an industrial smokestack, he explained that the six months spent researching the program was "a frightening experience":

> I don't smoke myself, but I now know that I'm getting the equivalent of two packs a day right out of the air. I'm inhaling a cup-full of dirt plus poison. I didn't know what emphysema was and perhaps you don't either, but you will. It's becoming one of the major killers. In fact, lung diseases as a whole are now the number one killer in Canada, and it's rather frightening to realize that most of our hospitals are in polluted areas. There are doctors who won't operate on dirty days. The density of automobiles in Toronto is four times what it is in Los Angeles. I used to think that air pollution was something they had in other countries, but we have it here and now in Canada, and you begin to feel like a fish in a poisoned pond.

After this dramatic opening, the film surveyed the wide range of air pollution problems in major centres across Canada and the United States.

It revealed that Canadian cities such as Toronto, Montreal, and Windsor had air quality equivalent to well-known polluted cities in the United States, such as Detroit, Chicago, and Los Angeles. The relationship between Sarnia's highly polluting oil and petrochemical industries and local physicians' reluctance to speak out against the effects these were having on locals' health was addressed. Industry representatives were interviewed, such as Dr. L.P. Roy of the Laval Industrial Association, who defended industry's right to self-regulate its emissions, while Jean Marier of Montreal's Air Pollution Control argued that the problem could be resolved only if "handled by public representatives." The film also included an interview with Hazel Henderson of New York City's 24,000-member Citizens for Clean Air. Speaking on her organization's efforts to secure clean air legislation, Henderson explained that "we have made air pollution a household word in New York City" and that, as a result, "nobody dared be against clean air."

The documentary switched gears thirty-three minutes in, turning the spotlight on the situation in Dunnville. Over a montage of farmers handling shrivelled produce and their cattle limping through fields, Burke dramatically summarized the issue:

> They noticed it first in 1961, again in '62 – worse each year. Plants that didn't burn were dwarfed – grain yields cut in half. He [a local farmer]'ll show you his fruit trees. The twenty-year-old orchard, trees that produced so richly for so many years. Now for six years, they've given up no fruit at all for market; random apples not worth picking. Finally a greater disaster revealed the source of the trouble. A plume from a silver stack – once the symbol of Dunnville's progress – spreading for miles around: poison. Fluorine. It was identified by veterinarians. There was no doubt. What happened to the cattle was unmistakable, and it broke the farmers' hearts. Fluorosis – swollen joints, falling teeth, pain – until cattle lie down and die, hundreds of them. The cause: fluorine poison from the air. Under arbitration, the Electric Reduction Company paid the farmers two hundred and eighteen thousand dollars for the loss of crops and cattle. Shriveled crops, limping cattle – but now is there a graver development?

This "graver development" was the suspicion that the fluorine pollution was causing human health problems. Burke was shown chatting with farmers Joe Casina and Ted Boorsma, who attributed their undiagnosed ailments, characterized by severely aching joints and swollen feet, to ERCO's effluent.

FIGURE 2   Canadian Broadcasting Corporation newsman Stanley Burke (grey suit) is seen interviewing dairy farmer Ted Boorsma during the filming of *The Air of Death*. Courtesy of Pollution Probe.

Then came the documentary's final, most contentious segment. Burke introduced Dr. Matthew Dymond, the Ontario minister of health, who was in studio for an interview. Burke announced that ERCO had declined to send a representative; instead, the set featured an empty chair. Dymond expressed concern regarding the human health problems portrayed but was quick to defend ERCO, stating that their pollution control efforts had limited "at least ... 90 percent of the emissions." Following up on the human health concern, a video then introduced Dr. Waldbott, who announced that two of the nine local farmers he examined displayed symptoms typical of those suffering from fluorine intoxication. Asked what he expected would happen if these two were left untreated, Waldbott's response was unequivocal: "If they continue to live in this area, eventually they are going to get more serious harm, serious damage to their joints – to their internal organs, particularly to their kidneys, and also to their brain and to the spine, which eventually will lead to death." When the documentary returned to the studio, Burke asked Dymond for his response. After acknowledging "that Dr. Waldbott has done a very great deal of work in the study of fluorosis" and that he was "among the most extensively quoted [authorities] on the continent and maybe in the world," Dymond emphasized that the symptoms were likely the result of a more common ailment, such as arthritis.

Discussion then turned to the jurisdiction of air pollution control. Dymond placed the onus on the federal government, noting that "air pollution doesn't recognize any geographic boundaries." A clip was shown of Allan MacEachen, the federal minister of health, who argued that the British North America Act assigns responsibility for addressing air pollution to the provinces. While he acknowledged that the federal government could play a role coordinating the provinces, MacEachen concluded by stating that "we do not have fresh plans at the present time for presentation to the provinces." As images of industrial smokestacks filled the screen, Burke delivered his stirring conclusion:

> So who will control air pollution? The cities? It's been tried and it hasn't worked very well. Among other things cities compete with one another to try to attract polluting industries. The provinces? Of course, but even provinces compete for industry and it's going on right now. Most authorities agree that it must be a cooperative effort from the federal government right on down, and most agree that it's urgent. We don't even have the detailed statistics in Canada. We don't know what's going on, and we may be right now well on our way toward our first disaster. We've cited some examples in this program and we could cite others, many others. Out on the prairies, "where the skies are not cloudy all day," they have fairly serious pollution problems. Jasper, up in the Rockies, is polluted. Banff could become polluted. Vancouver could have another Los Angeles situation, and experience elsewhere has shown that air can be cleaned up. I've driven through Germany, the industrial heartland of Europe, and the air is clear. Russia has imposed the highest standards of purity in the world. But in our society not much happens until the average citizen demands it.

The Response to *The Air of Death*

The Air of Death was a ratings success. According to a study by the CBC's Research Department, 16 percent of English-speaking Canadians over the age of twelve – or 1.5 million people – watched the documentary. This was considered an amazing achievement for an internal production. While the program attracted a steady audience across the demographics, the report's authors noted that 12 percent of viewers were teenagers, making it "an audience much younger than that normally attracted to most CBC information and public affairs programs." The film received "an overall index of enjoyment of 81," which, the authors noted, "represents a very high level of praise indeed," while "90 percent reported feeling that they knew either 'a great deal more' or 'quite a bit more' about the problems and dangers of air pollution than they knew before" as a result of viewing it.[22] Arthur Laird, Director of Research at the CBC, wrote to Murray Creed: "Actually, 'Air of Death' was *so* well received that it is difficult to point to anything in the program that, from the audience's point of view, went seriously wrong – nor to anything that, had it been done otherwise, would have been likely to increase substantially the program's general impact."[23]

The program also proved to be a critical success. According to Roy Shields's 23 October "TV Tonight" column in the Toronto *Star*, "Today we all feel a little more grimy thanks to Stanley Burke, producer Larry Gosnell and the boys of the CBC's farm department." As he explained: "This was a well-researched, highly-documented program that must have shocked thousands of easy-breathing viewers from coast to coast. For taking a firm journalistic position that Canadians have been living in a fool's paradise of pollution, the program did the nation a service."[24] Bob Blackburn, television critic at the Toronto *Telegram,* was equally enthusiastic about the production, calling it "one of the more venturesome things the CBC has done in public affairs." He was particularly taken by how the message was delivered: "It didn't get hysterical. It didn't have to. It just calmly recounted the manner in which not only city-dwellers but some rural folk also are quietly being poisoned while no one does anything effective about it." Blackburn posited that, if anything, the documentary was not sufficiently alarmist to jolt the public into action.[25]

The fallout from the documentary began on the night of the press screening – 19 October – when Ontario's health minister, Matthew Dymond, announced that his department would conduct medical tests to determine the source of the farmers' illnesses.[26] Eight days later, he announced a public inquiry into all forms of fluoride pollution in the Dunnville area that would explore its impact on human, animal, and plant health as well as its financial toll. Although the government conceded that the fluorosis poisoning found in local cattle was the result of ingesting "crops exposed to fluoride emissions," it argued that it was far less likely that there were any cases of human fluorosis, as only a small part of the human diet would consist of local produce, and even this was routinely washed and cooked prior to consumption.[27] For its part, ERCO maintained a steadfast public denial that its plant was causing human health problems, although ERCO board vice chair Omond Solandt expressed some concern about the company's culpability in a letter to Sir Owen Wansbrough-Jones, chair of the parent company, Albright and Wilson. Some residents collected and drank rainwater because of an unpleasant sulphur aroma in local wells, and Solandt noted: "It is highly unlikely but just possible that they could have ingested significant amounts of fluorine from this source."[28] Wansbrough-Jones asked Solandt, who also served as the University of Toronto's chancellor and as chair of the Science Council of Canada, to use his considerable influence to promote ERCO's side of the story behind the scenes.[29]

The commissioners charged with carrying out the provincial inquiry were announced on 6 November 1967. At the helm was Dr. George Edward Hall, who had recently retired as president at the University of Western Ontario. He was joined by Alex McKinney, a former president of the Ontario Federation of Agriculture, and Dr. William C. Winegard, president of the University of Guelph, was added in January. The choice of commissioners met with ERCO's approval, as Solandt was a long-time friend of Hall.[30] Not only were all three commissioners partisan Progressive Conservatives but, in the case of Hall and Winegard, they were also well connected with the fluoride industry. Earlier in the decade, Hall had served on the Morden Commission on municipal water fluoridation in Ontario; he later served as the honorary advisory director of the Health League, the foremost promoter of fluoride in Canada. Hall's appointment was opposed by Waldbott, who argued that a whitewash was in the offing, as well as by the local farmers, who unsuccessfully lobbied Dymond to select a new chair. Winegard, who later served as minister of science and technology in the Mulroney government, had served as editor of the *Canadian Metallurgical Quarterly* from 1965 to 1966, and had recently received an award from the Canadian Institute of Mining and Metallurgy for his contributions to the field of metallurgy. The farmers also opposed the selection of McKinney, claiming that despite his agricultural background, his Tory partisanship meant that he would not represent their interests.[31]

The Hall Commission

The Pollution Inquiry Committee (the Hall Commission) began on 22 January and concluded on 21 March 1968. Much of the hearings hinged on the expertise provided by five health consultants. Besides sharing a pro-fluoridation stance, the consultants lacked experience in treating and diagnosing fluorosis. One expert hired for the inquiry was Dr. Patrick Lawther, director of the Air Pollution Laboratories of the Medical Research Council in London, England, who had recently made news headlines when he proclaimed at a pollution control conference in Toronto that "air pollution is a field which contains more cranks and psychopaths ... than any other field I could have stumbled upon." He also refused to link air pollution to health problems, noting that after thirteen years of studying the matter, "we have produced no unequivocal results."[32] These medical consultants consistently rejected the idea that ERCO's effluent was having a negative impact on the health of the local populace.

The commissioners also relied on a selective reading of scientific research. As they explained in their final report:

> This report will not contain a complete survey of the [scientific] literature; it is not the responsibility of the Commissioners to do so. Since there is, in general, major agreement on the results of experiments, surveys and special studies, certainly amongst the recognized and accepted scientists, the conclusions reached by such eminent workers have been taken as the basis for comparison of the evidence elicited at the Committee hearings, where comparisons were relevant.[33]

Consequently, studies that documented human fluorosis and other forms of industrial fluoride pollution were routinely excluded, and the case of Garrison, Montana, was never discussed during the hearings.[34]

Evidence of deleterious human health conditions caused by ERCO was also denied a proper hearing. Locals complained on the stand of ill-effects, including sore eyes, burnt lips, and respiratory problems caused by the industrial dust settling in the area. However, the commissioners blocked local physician Dr. F.D. Rigg from discussing the residents' symptoms, arguing alternately that it was inappropriate to discuss patients' symptoms in their absence and that the doctor was not qualified to diagnose fluorosis. The commissioners also prevented discussion of a report prepared by the Ontario Water Resources Commission (OWRC) in 1965 that revealed fluoride levels as high as 37.8 parts per million – far above the danger threshold of 2.4 parts per million. Efforts by the farmers' lawyer to discuss this were blocked, with the promise by the commission's lawyer that it would be discussed later, when an OWRC representative was available to interpret the test results. When the topic was finally re-addressed, the results were summarily discredited because one of the thirty samples was not properly labelled.[35]

Also missing from the Hall Commission were the figures central to the creation of *The Air of Death*. From the outset, the CBC took the position that it would not participate in the hearings, arguing that provincial commissions lack jurisdiction over federal agencies. Likewise, the CBC took a strong position in support of those involved in the production of the film, promising to appeal any efforts to subpoena witnesses.[36] Although no subpoenas were issued, the commissioners did pressure Gosnell to provide evidence in support of fourteen contentious statements made in the documentary. Although the CBC initially refused to respond – a letter from Marcel Munro, acting general manager of Network Broadcasting (English), reminded the commission's secretary that the CBC "is accountable to Parliament for the conduct of its affairs and the discharge of its responsibilities"[37] – the network eventually relented and prepared a detailed, seventy-one-page response.

Dr. Waldbott was also absent from the inquiry. He wrote the Hall Commission on 1 January 1968 to announce that he would appear, but stressed his need for additional time to prepare his documentation. In February, he contacted the commission's secretary in an effort to arrange an appearance. Despite receiving a letter of acknowledgment, he later insisted that the Hall Commission did not attempt to work him into the schedule. The commissioners dismissed this notion in their final report, stating that they had "extended an invitation to Dr. Waldbott to appear before the Committee," but that "he saw fit not to submit himself for cross-examination."[38] Instead, Waldbott submitted a detailed brief containing updated evidence on examinations of twenty locals, in which "10 presented definite evidence of fluorosis, [while] seven should be suspected of ill-effects from fluoride."[39] Although the Hall Commission report acknowledged receipt of this brief, it noted that "the Committee rejects many of the statements made by Dr. Waldbott in his brief and accepts the testimony of the physicians and other scientists received in evidence and referred to or quoted in the Committee's report."[40] In his absence, Waldbott was the target of much mudslinging. Despite being recognized by Dymond in *The Air of Death* as one of the leading authorities on fluorosis, he was depicted throughout the hearings as a fanatical and irrational opponent of the fluoride industry.

The Hall Commission's report was tabled in the provincial legislature on 10 December 1968. Although some criticism was levelled at ERCO – particularly that it should "install the necessary equipment and modify their operations to reduce dust emissions from the lagoons, and emissions from the curing sheds, to acceptable limits under full plant operation" – the company was portrayed as a good corporate citizen that was "generous, and, in some instances, more than generous" when compensating local farmers. The commission conceded that ERCO was causing some damage to the surrounding agricultural economy, but it insisted that the "people of the Port Maitland area can be assured that there is no human health hazard associated with pollutants being emitted from the industrial plants in the area." The report directed considerable vitriol towards the CBC, stating that it "has no other alternative but to record that unwarranted, untruthful, and irresponsible statements were made by the publicly-owned and publicly-financed Corporation, the CBC. They treated a complex problem in a way designed to create alarm and fear. Their treatment was not in keeping with the standards which the public is entitled to expect from the Corporation." Furthermore, while the CBC program referred to the affected farmers as Dunnville residents, in actuality they resided in the neighbouring community of Port Maitland. Given that the residents of Dunnville would suffer financial losses as a result of this mistake, the commission recommended that they undertake legal action against the CBC.[41]

Not surprisingly, the report's findings drew support from ERCO. Solandt wrote Hall, noting that "I have watched your pollution investigation from the sidelines because I did not want to have an unfriendly press seize on our longstanding friendship. However, now that the Report is out and I have read it, I feel that I can safely write to congratulate you on doing an excellent job."[42] While media outlets generally accepted the findings of the Hall Commission report at face value, letters critical of it were published in the Toronto *Star* and the *Globe and Mail* in the ensuing days. Most notable was one by Gavin Henderson printed on 27 February 1969. The first executive director of the Conservation Council of Ontario and a co-founder of the National and Provincial Parks Association of Canada, Henderson wrote of "a disquieting similarity between the efforts to denounce Rachel Carson," the American author whose bestselling exposé of the detrimental effects of synthetic chemicals, *Silent Spring*, sparked a vicious backlash from industry, and the attempt to stifle environmental concern in Canada.[43] Comparisons between the Dunnville situation and *Silent Spring* were also made in the *Family Herald*, which ran an editorial on 26 October 1967 titled "How Many Dunnvilles to a Silent Spring?"

Furthermore, a wide range of supporters, including many prominent scientists, wrote the embattled CBC staffers following the tabling of the report. Dr. J.M. Anderson, secretary-treasurer of the Canadian Society of Zoologists and director of the Fisheries Research Board of Canada Biological Station in Saint Andrews, New Brunswick, wrote that "the film was a thoughtful, imaginative, and serious treatment of a problem well-deserving of widespread public attention ... Those associated with it are to be commended."[44] Dr. Henry Regier, associate professor of zoology at the University of Toronto, stated that "the CBC should be congratulated and honoured for this production when it is considered in a broad scientific ecological viewpoint."[45] Staffers also received a letter from Dr. Donald Chant, chair of the Department of Zoology at the University of Toronto and one of the resource people utilized during the making of *The Air of Death*. After briefly outlining the scientific shortcomings of the Hall Commission, including the failure to conduct bone biopsies that would conclusively determine if there were any cases of human fluorosis, he added that "the Commission's chapter on the CBC seems petulant, almost as if it resented your intrusion into its private preserve, and contains questions out of context from 'Air of Death.'"[46]

The CRTC Hearing

On 18 December 1968, just eight days after the Hall Commission report was tabled, the CRTC announced its intent to hold hearings on the subject. The notice of public hearing established a mandate to determine whether the CBC had acted responsibly in producing the documentary.[47] The hearing was not established to explore air pollution, and did not allow for "the introduction of evidence, scientific or otherwise of matters arising since the date of broadcast of the program."[48] These terms proved somewhat disappointing to those involved in *The Air of Death*, as they had hoped for an opportunity to address the misrepresentations made during the Hall Commission.

While the CBC continued to support its embattled employees, recognition that their interests were not entirely congruent led the corporation to hire Creed, Gosnell, and Burke their own separate legal counsel, in the person of Joseph Sedgwick, a prominent Toronto lawyer who had served as treasurer of the Law Society of Upper Canada in 1962-63.[49] The trio also began strategizing with Victor Yannacone, the renowned co-founder of the US-based Environmental Defense Fund. In these sessions, which involved numerous telephone calls and at least one weekend meeting,

Yannacone peppered the Canadians with advice. Hailing theirs as "the most worthy cause we have had in a long time," Yannacone emphasized the necessity of having all relevant research and documentation clearly organized and readily available during the hearing.[50]

The CRTC hearing began on 18 March 1969. The commission was headed by CRTC vice chair Harry J. Boyle. He was joined by Réal Therrien, a member of the CRTC's Executive Committee, and Dr. Northrop Frye, the noted literary critic and theorist. The commission began with a screening of *The Air of Death*. Before the first witness could take the stand, Jacques Alleyn, the CBC's general counsel, outlined the corporation's feelings regarding the hearing. He argued that the CBC required an untrammelled press, free from pressures other than those resulting from law. According to Alleyn, "this is the price to be paid for democracy."[51]

The first witness to provide testimony was Eugene Hallman, who discussed the chain of command, job responsibilities, and general broadcasting policies at the corporation. When Gosnell took the stand next, the CBC's strategy became apparent. After a brief discussion of the origins and development of the project, Gosnell spent the bulk of the next two days introducing the extensive research behind *The Air of Death* into the official record. With three filing cabinets of documentation and a list of approximately 170 research and production contacts at Gosnell's side, this was a move clearly intended to counter the Hall Commission's allegations of shoddy preparation on the part of the CBC. The approach worked. As Boyle announced partway through the second day of testimony: "If it is a matter of establishing the amount of research that Mr. Gosnell has undertaken with a crew in terms of his actual program, he has demonstrated now that I don't know how he had time for the program ... I would suggest to you that you have amply demonstrated this point – the degree and the extent of the research of Mr. Gosnell and his group. If it is possible to expedite it by filing it in a group, we would appreciate it."[52]

Gosnell was followed on the stand by Stanley Burke, who described his role in the production. Asked by Alan Golden, counsel for the inquiry, whether he felt that the subject matter justified exaggeration on behalf of the filmmakers, Burke assured him that "I don't consider that there was any exaggeration in the 'Air of Death' program. I think it was understated."[53] On 20 March, P.B.C. Pepper, counsel for ERCO, took the stand. He alleged that *The Air of Death* featured material emanating from Dr. Waldbott, "who some people might say was a crank ... who was emotionally committed, a propagandist for a cause."[54] Pepper concluded by arguing that *The Air of Death* must be held to a higher standard of factuality because of Burke's role as a prominent newscaster.

Larry Gosnell's appearance on the stand drew rave reviews from his superiors at the CBC. George F. Davidson, the corporation's president, wrote in a 31 March 1969 letter: "You made all of us proud, – all of us who belong to and believe in the CBC, – by the quality of your testimony and by the evident integrity reflected by your presence and your evidence given from the witness box."[55] This was followed by a letter from Eugene Hallman on 1 April, noting that "I admired the way you conducted yourself during the CRTC hearings into 'Air of Death.' The Corporation could not have had a better witness and I was proud of the way in which the research data had been assembled so carefully, not simply for the presentation at the hearings but for the broadcast itself."[56] Gosnell's performance was even more impressive in light of the fact that he was a last-minute replacement for Murray Creed, whose appearance at the CRTC hearings was cancelled two days earlier by the onset of labyrinthitis, an inner-ear disorder that causes hearing loss and balance problems.[57]

The CRTC Report

The CRTC released its report on 9 July 1970. *The Air of Death* received a general vindication, with the CRTC stating that "the program adequately reflected the information reasonably available at the time of the broadcast and is well able to stand as an example of informational programming backed by a wealth of research and serving a useful purpose." Furthermore, "it is the opinion of the Committee that Air of Death [sic] may well have been one of the most thoroughly researched programs in the history of television broadcasting." The report also noted "that the use of the term 'Dunnville' to describe the area allegedly affected by fluoride emissions was reasonable and proper in this instance." The production did not escape critique, however. First, the report argued that *The Air of Death* should have highlighted the fact that conflicting medical opinion existed regarding human fluorosis. The fact that the information broadcast was based primarily on the opinion of Waldbott, who was "known to hold sharply critical views on the effect of any fluoride emissions upon human health," should have been explained, as should the fact that his opinions were highly controversial within the medical community. Second, the report argued that the segment featuring Allan MacEachen wrongly implied that the federal government was powerless to address air pollution, as unaired portions of his interview indicated that the federal government was engaged in extensive research on the subject and was trying to coordinate the provinces in an effort to address the problem. In light of this, the report stated that "constructive statements should be given due prominence."[58] The report also criticized the fact that Dymond had commented on-screen about MacEachen's statements but that MacEachen had not been given the opportunity to rebut.

Despite the criticism, the CRTC report was viewed positively by the embattled CBC employees. "All in all I was very happy with the C.R.T.C. findings," wrote Creed in a 15 July 1970 memo to the CBC's regional supervisors. "There are things with which one could quibble but there seems to be little point in argument. Better than 'irresponsible, unwarranted and untrue' in any case." As Creed added, in the last line of the memo, "I believe we can now write Q.E.D. to Air of Death [sic]."[59]

The Birth of Toronto's Environmental Movement

The Air of Death became a key event in the formulation of broadcasting standards in Canada, particularly with respect to the creation of "balance of views" requirements by the CRTC.[60] More importantly, the warning contained in the documentary and the public efforts to discredit those responsible for its production inspired the creation of Toronto's first ENGOs, GASP and Pollution Probe. GASP was started by a cadre of Toronto's professional elite, including James Bacque, chief editor at Macmillan Company of Canada, Toronto City Council alderman Tony O'Donohue, a noted critic of pollution, and Dr. Alfred Bernhart, professor of civil engineering at the University of Toronto.[61] Bacque was alarmed by *The Air of Death*'s discussion of urban air pollution, and phoned Stanley Burke at the CBC headquarters with the idea of forming a citizen-based group to address the problem. "Stanley was quite welcoming and cooperative and he offered to help where he could," Bacque explains, "but he's not an organizational type." Recruiting was a simple matter: "When we started phoning around ... everybody that we contacted was in favor of doing something because they'd been alerted by that show." The initial meeting, devoted to organizational matters, was held at the home of Joseph Sheard, a prominent city lawyer. According to Bacque, it was a smoke-filled affair: "In our first meeting in Joe's living room, he was sucking on a pipe and so was I, and probably about a third of the people in the room were poisoning themselves with cigarettes. *[laughs]* We did notice the irony."[62]

The group was initially known by the rather formal name of the Citizen's Committee for Clean Air in Metro. This was changed to the more emotionally driven and memorable name Group Action to Stop Pollution, or GASP, prior to its public launch. As O'Donohue explained to the press, their goal was "to badger these governments who are dragging their feet on air pollution" and to "name names and demand action" against polluters.[63] The group made its public debut during the Ontario Pollution Control Conference in Toronto on 4-6 December 1967. Ordered by Premier John Robarts, the conference aimed to "provide a comprehensive approach to environmental pollution in all its aspects," including "the problems of air, soil and water pollution in agricultural, industrial and municipal contexts."[64] Given the context of the times – *The Air of Death* had been televised just weeks earlier and the Hall Commission was preparing to launch its investigation of the broadcast in January – the conference received considerable attention from the local press. GASP piggybacked on this media convergence. It circulated pamphlets advertising its first public meeting, scheduled two days after the conference's conclusion, and drew further attention by sending members wearing surgical masks to canvass commuters at the Yonge and Eglinton subway station. The three-hour public meeting drew an estimated crowd of three hundred.[65] The moderator was *Air of Death* host Stanley Burke, who opened the meeting by greeting his "fellow cranks and crackpots" – a clear dig at Patrick Lawther, a speaker at the Ontario Pollution Control Conference who, just days earlier, had dismissed pollution control advocates as "cranks and psychopaths."[66] The special guest speaker was Hazel Henderson, who spoke on the experience of New York City's Citizens for Clean Air, the group she co-founded in an effort to raise public awareness of air pollution and secure government legislation. Although government and industry were slow to recognize her group, Henderson urged those gathered to persist, noting that "there's simply no way to halt the public demand for the right to breathe." Matthew Dymond sent a note stating that he could not attend and that his Air Pollution Control Service officers were likewise unavailable – news that inspired heckling from the crowd. Also discussed during the meeting was the cost of improving Metro's air quality – which Alfred Bernhart pegged at $540 million, or $300 million if buildings transitioned to using natural gas – and future GASP activities, such as an Easter "breathe-in."[67]

A twenty-member permanent committee was established. Aside from Bacque, O'Donohue, and Bernhart, notable members included Gosnell, Burke, Margaret Scrivener, whose efforts to prevent development of the

FIGURE 3 Members of Group Action to Stop Pollution are shown promoting the organization's first meeting at the Eglinton and Yonge subway station, December 1967.
Source: Toronto *Telegram* fonds, ASC17107, 1974-002/091, Clara Thomas Archives and Special Collections, York University.

Toronto ravines system during the 1950s had caused her to be dubbed "the lady of the ravines," and Aird Lewis, a corporate lawyer who had, among other conservation initiatives, co-founded the Nature Conservancy of Canada in 1962. O'Donohue promptly stepped down after the meeting, informing the press that his status as a city alderman was inconsistent with GASP's need to remain politically nonpartisan.[68]

GASP's first major activity was a press conference on 25 January 1968 in which it "deplore[d] the atmosphere of recrimination, distrust and abuse" at the Hall Commission. Bacque, identified as spokesman of the four-hundred-member organization, accused the Hall inquiry of hiding important medical information from public scrutiny. Two residents of Port Maitland had been hospitalized for extensive testing for fluoride poisoning, with the provincial government picking up the tab. "If there is no evidence of fluorosis poisoning," Bacque asked, "why have they been kept in hospital for two months?" He also accused the commissioners of ignoring available medical experts.[69] The following day, GASP was officially established as a charitable corporation without share capital. The charter, signed by GASP's new directors – Bacque, Bernhart, Sheard, John Hunter Lytle, and Richard Alan Mansfield – described the organization as having an educational emphasis, designed to "educate and inform the general public, particularly of the Province of Ontario, of the health, aesthetic and economic effects of the air, water and soil pollution and the many technological and legal tools presently available for control."⁷⁰

At this point, GASP appeared to be on solid ground. It had a team of five directors and a twenty-member permanent committee. It had held a high-profile founding meeting and a widely covered press conference. On the other hand, the group had yet to make good on earlier plans, such as establishing a newsletter and holding protest actions. These shortcomings are attributable to the ENGO's status as a part-time pursuit of busy professionals. Moreover, despite having what appeared to be an impressive leadership group in place, membership on the permanent committee was largely honorific, with little or no actual involvement in GASP's activities. The burden of operations fell on a small core of dedicated volunteers. When Bacque, the most active of GASP's volunteers, left in the summer of 1968 for a year in France, the group fell into dormancy.⁷¹

GASP was given a second lease on life in January 1969, when it was announced that Tony O'Donohue would return and assume the newly created position of full-time executive director. He noted that in order to accept this unpaid position, he would abandon his job as an engineer and live off his salary as an alderman. Determined to run an advertising campaign highlighting the dangers of air pollution, he also announced plans for a fundraising drive to cover the necessary costs. In bringing O'Donohue back into the fold, the ENGO replaced one problem with another. It is true that by taking on the full-time position, O'Donohue offered the potential of an organized group that could proceed with fundraising and educational pursuits. On the other hand, when he left the organization in December 1967, he had cited his desire to avoid turning it into a partisan operation; now, however, he was looking ahead to a 1972 mayoral run, and it was inevitable that GASP would be seen as a platform for his political ambitions. The result was something of a Faustian bargain. GASP as an organization was revived and once again visible to the general public. Inexorably, however, it became politicized, which limited its long-term appeal to the wider populace.⁷²

The ENGO returned to action in March 1969, submitting a brief to the CRTC's investigation into *The Air of Death*. GASP's brief, presented by Bernhart and O'Donohue, summarized the deleterious effects of air pollution on human health and defended the film's more controversial statements. Calling *The Air of Death* "a very promising first step in making people aware of the filthy conditions of the environment in which we live," the brief also credited the documentary with mobilizing a previously dormant populace:

> It also helped this organization – Group Action to Stop Pollution – to become organized and increase in strength and attract more members. It was gratifying to see so many people not associated with pollution previously take a keen interest in combatting the legacy of polluted air that we have left and are continuing to leave to succeeding generations of Canadians. We, as ordinary citizens, have been able to band together for the one big purpose: and that is, to halt the expanding pollution of our environment. We feel that the CBC's program "The Air of Death" has been of tremendous value to us in recruiting the average citizen to our ranks and we would hate to see the CBC, or any of the news media, be they press or radio, muzzled and made the puppets of big business or political parties.

The brief also highlighted GASP's concern "that the whole battle against all types of pollution has been dragged through the credibility filter":

> The average citizen is witnessing a display that will only weaken the cause of air pollution control, because very obviously it is an open battle between a news media on one hand and the "powers that be" on the other side. The man in the middle is Joe Doe and Family, down the street, who will still have to live with their children in an increasingly polluted environment with not much hope of ever halting the deadly fumes that daily are pumped into our atmosphere.[73]

Another group inspired by *The Air of Death* was Pollution Probe. The roots of this group can be traced to the University of Toronto's student newspaper, *The Varsity*, whose staff was concerned that the documentary's warnings of environmental degradation were being overshadowed by efforts to discredit the filmmakers. Members of the staff spent their February study week investigating pollution levels in Toronto. A 24 February 1969 article by news editor Sherry Brydson announced that they were sufficiently concerned by their findings that they had formed "a group action committee, the U of T Pollution Probe." As the article explained, the nascent organization was mandated to investigate the origins and effects of pollution, as well as "mobilizing the public, private and government sectors to action in removing the poisons from our air – before it's too late."[74] Brydson wrote two follow-up articles. The first questioned the veracity of the Hall

Commission report in light of the fact that "the commissioners did not hear testimony from *a single doctor who had personally diagnosed or treated a case of flourosis* [sic]," while the second questioned the role of the University of Toronto in the Dunnville affair, highlighting the fact that its chancellor, Omond Solandt, was vice chair of the ERCO board.[75]

Brydson's message resonated with the university community. The first two meetings, held in February and March 1969, attracted hundreds of concerned students and faculty. In one respect, this was hardly surprising, given the powerful student movement at the University of Toronto. Many issues had been addressed and debated during this period, but two ultimately took precedence among the students: reforming university governance in an effort to increase their influence, and voicing opposition to the war in Vietnam.[76] Amid the teach-ins and protests of the period, students were politicized to a degree unimaginable in their parents' generation. This climate proved integral to the creation of Pollution Probe. Early member Stanley Zlotkin explains: "In the sixties we, the people at the university, really had the sense that (a) we had the obligation to move things along in the right direction and (b) we had the capacity to do it. It was a period of fairly non-passive thinking, and I think Pollution Probe was a manifestation to a certain extent of that. You know, we really did feel we could influence what happened in the future and it was ours to influence."[77] Just as important as *The Air of Death*'s alarming message in attracting support from the university community was the ensuing controversy. On why the documentary inspired so many to react, Brian Kelly, another early member of Pollution Probe, explains that "it was not just a story about industrial air pollution, it was a story about Canada's economic elite having the power to suppress that information ... It was a classic late-sixties struggle between the economic elites versus the public interest. It was an issue about power, not pollution necessarily."[78]

Another noteworthy feature of the upstart organization was that it was officially registered as a project of the Zoology Department. This resulted from a meeting between Brydson and department chair Dr. Donald Chant, whose assistance she sought in writing a brief to the CRTC. Chant, a native Torontonian, was an acarologist whose work focused on the use of mites and ticks as an alternative to pesticides in controlling insect populations harmful to forestry and agriculture. A second-generation academic – his father served as head of the Department of Psychology at the University of British Columbia – he grew up with a deep-rooted love for the natural world. As a child, he joined the Young Field Naturalists of

Toronto and took weekend nature classes at the Royal Ontario Museum. While working on his undergraduate and master's degrees, he held summer jobs researching fish species in the Bering Sea, as well as spruce budworm and mites in British Columbia. After obtaining his PhD from the University of London in 1956, he had a varied career: director of the Canada Department of Agriculture's Research Laboratory in Vineland, Ontario, from 1960 to 1964, and chair of the Department of Biological Control at the University of California, Riverside, from 1964 to 1967, before assuming his post at the University of Toronto.[79] Years later, he reflected on Brydson's request for help with the brief, and his decision to support the students' efforts to form an anti-pollution organization:

> I thought this was a fine idea, not only because I thought the program was fair but also because here was a group of young students who were concerned enough about, not Woodstock, or student power, or the Berkeley riots or whatever, but about the environment and its integrity to actually stand up, do some hard work, and be counted. It was for that reason that I did not turn them away or give only token support, but rather committed departmental resources and space to help them.[80]

This departmental support proved invaluable. In one respect, it provided them with the physical infrastructure necessary to operate. Office space was provided, initially at 91 St. George Street and by September 1969 in the Ramsay Wright Zoological Laboratories. While this set-up was not always ideal – Pollution Probe members from this period recall working long hours amid Bunsen burners and other scientific equipment – it provided the group with a place to meet and do their work. More important, the affiliation provided Pollution Probe with instant credibility. From the outset, the organization emphasized the need to back its activities with sound science; otherwise, member Rob Mills noted in the 1 April 1969 newsletter, "we are reduced to the status of a howling pressure group."[81] While Chant would remain Pollution Probe's most vocal champion, providing them with the necessary support and often serving in the early days as a public spokesperson and adviser, he was by no means their only ally within the Department of Zoology. In fact, the department was rife with activist-oriented faculty who were willing to lend their expertise. In October 1968, Dr. Henry Regier, a limnologist, and Dr. J. Bruce Falls, a behavioural ecologist, organized an International Teach-in on campus devoted to issues related to human population growth. The event attracted over three thousand participants and spawned the edited collection *Exploding Humanity: The Crisis of Numbers*. Dr. Chris Plowright, an entomologist who co-founded Toronto Zero Population Growth (TZPG) in March 1970, and Dr. Ralph Brinkhurst, a specialist in aquatic worms, were also notable early supporters.[82]

Within a month, Pollution Probe had attracted over 140 members. Its first action was the development of its CRTC brief, which stressed the importance of allowing the CBC to continue its "public education" productions unencumbered, noting that "many of us [Pollution Probe members] were not previously aware of the seriousness and complexities of air pollution problems." Stating that the documentary was factual and not overly sensational, the brief also raised the fear that a contrary verdict by the CRTC would dissuade the CBC from producing comparable, much-needed programming. Pollution Probe's brief also emphasized the need to deal with the situation in Dunnville: "It is evident that there is some sort of health problem in Dunnville, and although we are not 100 percent positive about the source of the problem, it nonetheless seems obvious to us that Dunnville is still in trouble ... This is a frightening and serious reality and we hope the CRTC will not forget this."[83] The brief was well received and Pollution Probe was invited to appear before the commission, an opportunity extended to just four other interested parties.[84]

Conclusion

When Larry Gosnell first envisioned *The Air of Death*, his aim was to raise public awareness of the widespread problem of air pollution. In attracting 1.5 million viewers, it can be safely surmised that he succeeded in this mission. Despite popular and critical acclaim, the program's harsh depiction of industry's willful contribution to the problem resulted in a campaign to discredit those involved with its production. This led to nearly two years of anxiety and uncertainty for Gosnell, his colleagues, and their families, but in the end the filmmakers were exonerated by the CRTC.

At the conclusion of *The Air of Death*, Stanley Burke announced that "not much happens until the average citizen demands it." If this was intended as a challenge, it was one duly met by those who created and filled the ranks of GASP and Pollution Probe. Although some were moved to act on realizing the severity of the air pollution problem, the founding of Pollution Probe reveals that others responded to the persecution of its messengers. In this sense, the embattled filmmakers came to represent something more than the story they covered – the suppression of the public good by members of the corporate community. Thus, ERCO's efforts to discredit those involved with the CBC production had major repercussions, as *The Air of Death* and the attendant controversy inspired the formation of Toronto's first ENGOs.

2
The Emergence of Pollution Probe

IN MAY AND JUNE 1969, a number of lifeless mallards were found off the shores of the Toronto Islands. Their deaths, attributed to the spraying of the pesticide diazinon by Metro Toronto Parks Department employees, were seized on by Group Action to Stop Pollution (GASP) and Pollution Probe as an example of society's careless use of deadly chemicals. To raise public awareness of the dangers of diazinon, the ENGOs organized a public inquiry concerning the dead ducks. The inquiry featured a panel of distinguished commissioners, including Dr. Marshall McLuhan. It confirmed diazinon as the cause of the deaths and recommended that use of the pesticide be severely curtailed.[1]

Just days later, on 18 July, Pollution Probe received an early morning tip that the Metro Toronto Parks Department was again spraying trees on the Toronto Islands with diazinon. At 6:00 a.m., Pollution Probe's Tony Barrett hurried to the islands aboard a rented watercraft in order to capture samples of the chemical. Dismayed by the Parks Department's continued use of diazinon when less lethal alternatives existed, GASP and Pollution Probe filed a writ of mandamus in the Supreme Court of Ontario, asking the court to prohibit the Parks Department from using the chemical. Although the writ was rejected, the ENGOs' campaign against diazinon ultimately resulted in a Royal Commission that addressed what killed the ducks and the utilization of synthetic chemicals in Ontario. The campaign against diazinon did not proceed entirely as planned, but it marked the first foray of GASP and Pollution Probe into the world of high-profile environmental activism. It would not be their last.[2]

Despite a common catalyst – *The Air of Death* – the overlapping histories of GASP and Pollution Probe offer a study in contrasts. GASP was the part-time pursuit of members of the city's professional class. Although it enjoyed an enthusiastic inception, the group lumbered along for two and a half years before going defunct in the summer of 1970. Pollution Probe, primarily composed of university students, gained a popular following after its February 1969 launch. Unlike GASP, it thrived and developed into one of the leaders of Canada's early environmental movement.

This chapter will examine the history of Toronto's original ENGOs through the summer of 1970. During this period, the benefits of Pollution Probe's affiliation with the University of Toronto's Department of Zoology became apparent, as it provided the ENGO with resources to pursue its activities full-time. Support from the department did not guarantee success, as demonstrated by the lacklustre emergence of Toronto Zero Population Growth (TZPG) in the spring of 1970. Rather, Pollution Probe's success was the result of its institutional support, the involvement of a dedicated core membership, canny leadership, and its ability to tap into Toronto's business community.

Organizing Pollution Probe

Within just a month of launching, Pollution Probe had delivered a brief to the CRTC but it still lacked an organizational structure. Initial plans to choose an executive via mail ballot, with each member receiving a single vote, were abandoned when it became apparent that the group was destined to be overtaken by hippies and more radical elements that approved of violence against polluters.[3] John Coombs, who was attracted to the early meetings because of his friendship with fellow Upper Canada College (UCC) alumnus Tony Barrett, recalls prompting the decision to appoint a temporary executive at the sparsely attended meeting on 17 March 1969:

> I remember Don Chant looking very disconcerted and frustrated that they weren't going to get this thing, Pollution Probe, going the way they wanted, so on an impulse I just got up and a bit like an impromptu dictator said, "Well, we don't have time for elections, we're just going to have to appoint people." I just sort of said, "Who would volunteer to do this role, who would volunteer to do that role," pretending that I didn't know anyone there, that I was just sort of taking over as [an impartial] group moderator.[4]

Consequently, Coombs's UCC colleagues Rob Mills and Tony Barrett were appointed President and Vice President (Administration), respectively, while Geoff Mains was appointed Vice President (Research), and Sandra Woodruff became Vice President (Communications). In the 1 April 1969 edition of the *Probe Newsletter*, the group excused its actions by explaining that there was a consensus among the thirty-five in attendance that it was necessary to "elect" a temporary government to carry it through exams and the summer. "While not an entirely democratic move," it was reasoned "that with the membership at 150 and growing every day, it would be impossible at this time to do anything else." Follow-up elections, as well as a potential re-structuring of the executive, were slated for September.[5]

The structure of Pollution Probe's provisional government demonstrates that initial plans called for the group to follow the model of traditional campus clubs. It was designed to be a part-time student operation that investigated pollution in the city and drew attention to the problems. In an ironic twist, given previous efforts to root out hippie elements, a dramatic restructuring occurred in autumn 1969: executive positions were abandoned in favour of a flat organizational structure. Monte Hummel, who joined Pollution Probe shortly after the organizational makeover, noted that "we had a very egalitarian group; the process was as important as the goal; we had no hierarchies; we had no bosses. Titles were very sensitive, so we didn't have them; everybody was a 'co-ordinator,' and not a leader or president, or anything like that."[6] Further extending the principle of equality within the group, Pollution Probe's weekly meetings were run according to the belief that each member deserved an equal say. The meetings soon became notorious for seemingly never-ending debates over matters big and small, and the tendency to last for hours.[7]

Pollution Probe's membership consisted primarily of middle-class individuals, a reflection of the standard makeup of Canadian universities during this period. The common denominator, aside from obvious concern for the environment, was having summer camp and cottage experience while growing up. As Lynn Spink explained, "I think that direct connection to the land, to the water, to the environment, had something to do with the passion in which we wanted to save what we experienced directly."[8] Other members of Pollution Probe would end up spending their summers working in the many camps that had blossomed in postwar Ontario, particularly those in Haliburton and Muskoka, which catered, in the words of historian Sharon Wall, "to a well-to-do, upwardly mobile, middle-and-upper class clientele."[9]

A small group with elite connections played a pivotal role in shaping Pollution Probe, however. Sherry Brydson, whose articles for *The Varsity* were central in getting Pollution Probe off the ground, was the granddaughter of Roy Thomson, the 1st Baron of Fleet, founder of the Thomson Corporation, and Canada's wealthiest person. Tony Barrett, a popular and energetic commerce student credited with keeping Pollution Probe afloat during its early years, John Coombs, and Rob Mills were graduates of the prestigious Upper Canada College. According to Mills, not only did UCC impress on its students the need to take a leadership role, but growing up in an environment where friends and family were influential figures provided an ability "to find the cogs of power and influence society ... We were aware of where power came from and where money came from."[10] This core also helped defend Pollution Probe from disparaging critiques. As Dr. Ralph Brinkhurst, an early faculty supporter notes: "One of the impressive things was that they couldn't be dismissed as sort of hairy radicals because they were all so conformist looking. You know, tall, upright, white Anglo Saxon. Clean, short haircuts. *[laughs]* All of the right accents."[11] This fact did not go unnoticed by the members of Pollution Probe. According to Ann Rounthwaite, the daughter of a prominent London, Ontario, medical doctor, its establishment appearance enabled it "to be heard by the media as well as our target audience in a way that a group of hippie environmentalists wouldn't have been."[12] This set in place a unique characteristic of the organization. Although Pollution Probe regularly singled out and criticized companies that were harming the environment, it was also quick to seek allies and cooperation from within the business community.

Barrett's involvement was vital in the development of Pollution Probe. The son of an affluent Toronto advertiser, he graduated from UCC in 1964 and promptly enrolled at Trinity College, the smallest and most exclusive of the University of Toronto's federated colleges, where he studied commerce. His privileged upbringing was rounded out by his attendance at Camp Hurontario, first as a camper and later as a counsellor. Plans for a career on Bay Street changed, however, when Barrett read Brydson's call to arms in *The Varsity*. As Rob Mills explains, immediately thereafter "he just quit everything and spent full-time working on it [Pollution Probe]."[13] Blessed with a dynamic personality, he served as a magnet that drew old friends, such as Coombs, Mills, and Monte Hummel, into the Pollution Probe fold, while also attracting others. As Coombs noted in an interview, "I would not have done it [joined Pollution Probe] if Tony hadn't been involved."[14] Barrett was also known for his keen sense of humour. Often seen wearing a green military helmet while in the office, he used humour to break the ice when contacting business leaders. Peter Middleton recalls Barrett phoning Xerox's Canadian headquarters and mimicking a photocopier by duplicating everything he said. He also had the audacity to greet prominent business leaders and politicians by their given names. According to Middleton, "Some people, it made their day. They were amused, charmed. Others, they sort of said, 'Who is this freak?' But the folks who said 'Who is this freak?' would never have been attracted to do anything with Pollution Probe anyways, so there was little to be lost and much to be gained trying."[15] Aside from breaking the ice with potential supporters, Barrett's humour helped bring levity to the often gloomy work of environmental advocacy. His most important contribution to Pollution Probe, however, lay in helping to infuse the group with a business sensibility, which proved to be integral in its development from a student club to a high-profile ENGO.

One of Barrett's first organizational projects was the creation of a Board of Advisors. Plans for this body were first announced in April 1969, with the dual purpose of providing advice "on our approaches to and management of our projects" and "to carry back to their outside colleagues word of Our Cause and to express and seek support for Pollution Probe."[16] The initial version of the Board of Advisors was in place by September 1969 and included Drs. Chant (who served as chairman), Brinkhurst, and Regier of the Department of Zoology, Dr. R.W. Judy of the Department of Political Economy, Dr. Phil Jones of the Department of Chemistry, Dr. J.H. Dales of the Department of Economics, Dr. Marshall McLuhan of the Department of English, and CBC broadcaster Stanley Burke. The composition of the board, coupled with Dr. Chant's visible role in the ENGO's early history, led to some suspicion that the group was dominated by university faculty and their interests. This notion is sharply rejected today by Pollution Probe's early membership, who, while admitting the importance of having a reputable board for opening doors in the business and political worlds, note that it had no impact on their day-to-day activities and provided advice only when requested.[17]

Pollution Probe's affiliation with the university was not without controversy. There was clearly animosity between the ENGO and university chancellor Omond Solandt, as demonstrated during the June 1969 convocation ceremony. "As you walked across the stage he was the guy that shook your hand and gave you your degree," recalls Brian Kelly. "A number of us ... put a Pollution Probe button on our lapels and as we came up to him either pulled our gowns aside or turned our lapels over to flash the

Pollution Probe button at him."[18] Solandt, for his part, was known to have raised objections to Pollution Probe's university affiliation in meetings of the Board of Governors, which were ultimately parried by Chant and university president Claude Bissell.[19] Bissell's defence of the group was outlined in a 27 May 1970 speech. As he explained, Pollution Probe's affiliation did not mean that its actions were endorsed by the university but rather "that its work will be serious, with a scientific basis." Noting that the group would occasionally be wrong on issues, he nonetheless ended his speech by calling them "a happy harbinger of a saner world."[20] Bissell's support for Pollution Probe and its activities appears to be the result of two factors. First, he was an ardent backer of Donald Chant. Given Chant's intimate connection with the group, this support was naturally extended to Pollution Probe. Second, the group arose at a time of increasingly strident unrest among the university's student population. During the 1960s, university students throughout the western world were politicized to an extent never before seen. In some cases, this led to violence. While the University of Toronto managed to avoid the worst of this, Bissell was in constant fear that the protests on campus might escalate, particularly after Steven Langdon, president of the Students' Administrative Council, mused openly about increased tensions in the 1969-70 school year. In light of this, he was heartened to see these students dedicating themselves in a peaceful and generally orderly fashion to a constructive purpose.[21]

Although Pollution Probe drew its support from across the university community, active members with a science background were rare. This made the contribution of Brian Kelly all-important. A zoology student in the last year of his three-year degree at Scarborough College, Kelly joined Pollution Probe after reading Brydson's articles in *The Varsity*. As he recalls: "Being a zoology student in the late sixties, you couldn't help but be interested in environmental issues." This, combined with a predisposition to activism, made him a natural fit for the upstart group. At the time, Kelly had planned to continue his education, by first completing the fourth year of undergraduate studies at the University of Toronto's downtown campus, then pursuing a master's degree in California before returning to Scarborough College to pursue a PhD under Dr. Fred Urquhart, who was famous for his work on butterfly migration patterns. During the summer of 1969, he initiated his first action as part of Pollution Probe when he noticed raw sewage floating in the Highland Creek. Tracing the problem to the Cumber Street Pumping Station, Kelly contacted the Metro Toronto Public Works Department. Unhappy with their response, he contacted the local CTV affiliate, which featured the story in its nightly news. After enrolling at the downtown campus in September, his career plans began to shift: "I spent nearly all of my time working for Pollution Probe and frankly very little of my time attending classes. *[laughs]* So I withdrew, with the intention ... [that] I was going to take one year off to work for Pollution Probe and complete my makeup year and then go on with my academic career."[22] As it turned out, he remained with Pollution Probe until 1974.

THE DEAD DUCKS CONTROVERSY

Despite its growing presence on campus, it was not until a bizarre series of events involving the spraying of pesticides on the Toronto Islands that Pollution Probe began capturing the attention of the city's population. On 16 and 23 May 1969, William Carrick, a wildfowl expert with the Ontario Waterfowl Research Foundation, visited the islands to capture mallards for experimental purposes. To facilitate collection, Carrick baited food with alpha-chloralose, a narcotic used to immobilize birds. When consumed in heavy doses, alpha-chloralose is known to render mallards unconscious; in this case, it resulted in the drowning death of no fewer than twenty-seven ducks. On discovering numerous birds in different stages of paralysis, Robert Van der Hoop, superintendent of the Toronto Island Park, asked Carrick to vacate the premises. Van der Hoop, however, did not immediately inform his superiors of these events.[23]

Later that month, Algonquin Island resident Martin Sawma contacted the Metro Toronto Parks Department numerous times, inquiring about the pesticides that were being sprayed on the islands' trees. Parks employee Robert Siddall, unsure what chemicals were being used but frustrated by the repeated calls, picked one from a list of chemicals he saw posted near the telephone. He told the caller that they were spraying diazinon, an organophosphate pesticide that causes death through the overstimulation of neurotransmitters. As Siddall later explained, "it [diazinon] sounded like a good term so I told him that was it ... If it had been bicarbonate of soda [on the list] that is what I would have told him."[24] Confusion grew when Tommy Thompson, Toronto's superintendent of parks, impulsively announced that diazinon had indeed been sprayed on the islands. As he explained at the time, "Hell, when one of my men called and said he thought they should spray I told him that some birds might die and I also told him to go ahead ... It's either that or have the trees dying and people getting covered in slimy caterpillars when they visit The Islands."[25] The blustery superintendent retracted his story soon after, admitting that he

FIGURE 4   Outraged over the apparent pesticide-induced death of ducks off the Toronto Islands, Pollution Probe and the Group Action to Stop Pollution organized a public inquiry into the occurrence, July 1969. *Source:* Toronto *Telegram* fonds, ASC17146, 1974-002/161, Clara Thomas Archives and Special Collections, York University.

had never actually verified the chemical's usage, but by that point members of the city's environmentalist organizations sensed a cover-up in the works and Thompson was unable to convince anyone otherwise.

Events escalated in June. Eleven dead ducks were found in waters surrounding the Toronto Islands and were sent for analysis to the Department of Physiological Hygiene at the University of Toronto. The tests were assigned to a junior member of the department, Dr. Hubert Hughes. His test results revealed diazinon levels as high as 66 parts per million, which Dr. Chant would call "the highest level [of diazinon] ever recorded anywhere in the world."[26] These test results strengthened the environmentalists' belief that pesticides were being used recklessly in Toronto and that although only ducks were affected at the moment, they had the potential to endanger human health in the future. Frustrated with Thompson's flippant behaviour – he was quoted in the press as describing the affair as "a tempest in a teapot"[27] – and convinced that theirs was an open-and-shut case that deserved further publicity, the members of Pollution Probe and GASP decided to launch a public inquiry into the matter. Although the inquiry was officially co-sponsored by the two groups, GASP's participation was overshadowed by that of Pollution Probe, and of its members only O'Donohue and Bernhart were involved.

The public inquiry was held on 7-8 July at City Hall. Although lacking certain powers, such as the ability to subpoena witnesses and hear testimonies under oath, the two-day event benefited from the participation of three high-profile commissioners: Dr. Ernest Sirluck, Dean of Graduate Studies at the University of Toronto; Dr. Robert McClure, the Moderator of the United Church of Canada; and Dr. Marshall McLuhan, director of the University of Toronto's Centre for Culture and Technology. The Department of Physiological Hygiene's test results were presented as the central piece of evidence, alongside Thompson's earlier statements concerning the use of diazinon on the islands. Although Thompson initially announced that he would boycott the event, he appeared on the second day. Having reviewed the spraying records, he testified that a substantially safer pesticide, methoxychlor, not diazinon, had been used on the islands. The commissioners, suspicious of Thompson's changing story, came down on the side of the evidence provided by the Department of Physiological Hygiene's tests, and attributed the ducks' deaths to diazinon.[28]

The diazinon issue did not fade with the end of the public inquiry. In a 16 July story in the *Star*, Metro Toronto chairman William Allen described the inquiry as a "witch hunt," decrying it as "unauthorized and unqualified." Furthermore, it was revealed that although Mayor William Dennison and city controller Margaret Campbell had wanted Thompson to cooperate fully with the public inquiry, they were overruled by the Metro Executive Committee. On 18 July, Pollution Probe's Tony Barrett was tipped off by an island resident that the Parks Department was applying pesticide to trees on the islands. Barrett rented a boat and raced over. Catching Parks staff in the act of spraying, he took a sample of the pesticide they were applying; tests at the University of Toronto's School of Hygiene showed it was diazinon. Pollution Probe and GASP filed a writ of mandamus in the Supreme Court of Ontario, with an eye towards forcing provincial Health Minister Dr. Matthew Dymond to cancel the Metro Parks Department's licence to spray pesticides. Dymond instead requested that Metro voluntarily refrain from spraying diazinon until its use was reviewed by the Ontario Pesticides Advisory Board, a move that prompted the ENGOs to suspend their legal proceedings.[29]

The situation began to move towards resolution during the Pesticides Advisory Board hearings. Beginning in July, the board interviewed twenty-one people, including Dr. Hubert Hughes. Sufficient doubt was cast on the conclusions reached in the July public inquiry that on 2 September the board recommended "that a Committee of Inquiry be appointed to investigate the matter on a judicial basis."[30] This suggestion was endorsed by the provincial government and on 19 September, Dr. Martin Edwards, chair of the Department of Physics at the Royal Military College and president of the Federation of Ontario Naturalists, was appointed to head a Royal Commission on Pesticides. On 8-16 December in Toronto, the Edwards Inquiry focused on the validity of Hughes's test results. After testimony from Carrick and Hughes and a reanalysis of the available data, it became apparent that Hughes had botched the initial tests. He had failed not only to include a "blank" non-poisoned duck with which to compare the results but also to accurately measure the level of diazinon present. Edwards concluded: *"The only waterfowl whose deaths are definitely attributable to the use of pesticides on Toronto Island between April 1st and August 1st, 1969 died as a result of the use of alphachloralose."*[31]

The environmentalists were wrong in asserting that diazinon had been responsible for the death of the ducks, even though they correctly attributed those deaths to the careless use of chemicals. Fortunately for them, this fact was overlooked, and Pollution Probe continued to be a favorite of the press, an ironic development considering its emphasis on "sound science." Even more so, they tapped into pre-existing concerns that synthetic chemicals were deleterious to the environment when used carelessly. Edwards's report called for an end to the use of alpha-chloralose in capturing live ducks; he also recommended that the provincial and federal authorities pass an environmental quality act similar to the National Environmental Policy Act recently passed in the United States. According to Monte Hummel: "We were convinced it was the pesticides. They're poison, [they] spread poison, ducks died; *ergo*, diazinon killed ducks. It turned out not to be that at all. We lost the battle but we won the war ... Our tilting at windmills had actually raised our profile."[32]

Highway Litter and Non-Returnable Containers

Pollution Probe made its first foray into waste issues in August 1969. The group decided to tackle the problem in its most superficial and easy-to-remedy form: highway litter. Roadside cleanups had become a popular aspect of civic pride in the United States, largely as a result of the Keep America Beautiful public education campaign, which was funded by glass, steel, aluminum, and paper container manufacturers in a concerted effort to place the onus for maintaining an aesthetically pleasing environment on individuals. As Pollution Probe later acknowledged, although "the consequences of littering are hardly serious – relatively," it was "really a state of mind in the public, an attitude towards the environment which we tested."[33] On 1 August 1969, nine of its members gathered the soft drink bottles and cans found along a two-mile stretch of Highway 400 to the north of Highway 7, and along the Highway 400-Finch interchange. Over the course of ninety minutes, they filled ten potato sacks with roughly a thousand discarded beverage containers.

Eighteen days later, on 19 August, Pollution Probe held a press conference at Queen's Park. With the sacks of refuse emptied onto the legislature's front steps,[34] Tony Barrett, wearing his trademark plastic military helmet, declared that he and his colleagues "carried out this demonstration today in order to illustrate graphically and tangibility [sic] the dimensions of one aspect of the problem of pollution and to show that the cause and the remedy ultimately lie at the doorstep of the individual." He explained that highway litter was not just an "aesthetic burden" but also a financial one, as the Ontario Department of Highways spent $1 million in roadside cleaning each summer. He announced that, in an effort to reduce littering by motorists, Pollution Probe was launching a roadside monitoring project during the upcoming Labour Day weekend. Teams of five would be posted along selected stretches of highway. The teams would consist of two "spotters" to catch passers-by in the act of littering, a cameraman to photograph the offender's licence plate, a "fetcher" to retrieve the item of litter, and a secretary to record the pertinent information. "Once the drivers [sic] name has been obtained from the registry bureau," Barrett noted, "information will be sworn out against him and he will be required either to pay the fine on the summons forwarded to him ($5-$50) or contest the issue in the local magistrate's court."[35]

As hoped, the press conference garnered considerable media attention, including a same-day page 2 story in the Toronto *Star* and page 5 coverage in the next day's *Globe and Mail*. It also ignited a minor controversy in the *Globe and Mail*'s editorial pages when a condemnation of Pollution Probe's "plan to initiate a guerrilla police action" drew letters supporting the anti-litter campaign.[36] Pollution Probe was pleased with the awareness it raised but disappointed with subsequent events. Just days later, on 25 August, Harold Adamson, the Metro police force's deputy chief, was called before the Toronto buildings and development committee. Asked to enforce the existing anti-litter by-law, Adamson responded that the police were powerless unless the offender cooperated – an unlikely scenario. Furthermore, when Pollution Probe attempted to bring its first case to court in September, it was rejected by a justice of the peace. Nonetheless, the increased focus on the issue enabled Pollution Probe to form working relationships with the Department of Transport to reform the relevant legislation and the Department of Highways to find alternative solutions to the litter problem.[37]

Emblematic of the highway litter problem was the increasing use of non-returnable soft drink cans. Introduced in Canada in the mid-1950s, cans were not popular initially because they tended to give drinks a "tinny" flavour. It was only after glass manufacturers introduced the disposable bottle in the 1960s that non-returnable containers began to gain significant market share. Promoted as a convenience item, in 1970 these accounted for 35 percent by volume of all soft drinks sold but resulted in twice as much waste as returnables. As if this was not reason enough for concern, announcements that leading soft drink brands Pepsi and Coca-Cola would soon begin using non-returnable plastic containers showed that convenience containers were likely to continue increasing their market share.[38]

Pollution Probe issued its first public denunciation of non-returnable containers on 19 August 1969, in conjunction with its Queen's Park press conference. In a separate press release that day, the organization made it clear that it held the soft drink industry responsible for "encouraging an unnecessarily wasteful and polluting packaging system by promoting soft drinks in cans and non-returnable bottles." Noting that each returnable container was used ten times, the press release explained that this made it a more cost-efficient choice for manufacturers and consumers, costing $1.13 per gross of ten-ounce bottles, compared with between $3.40 and $4.60 for the equivalent ten-ounce non-returnable bottles. The press release concluded: "When there is an alternative to such a wasteful and costly system, an alternative that would cut the garbage from one industry alone down to 10 percent of its present rate, surely we should do all possible to see that the alternative is followed."[39] As would become apparent, however, the driving force behind the growth of non-returnable soft drink containers was the retailers. Returnable containers required retailers to provide storage space and incur additional handling costs, whereas non-returnables eliminated both concerns. This perspective was made abundantly clear in a letter from the president of the Mac's Milk Ltd. [now Mac's Convenience Stores] chain to Pollution Probe, noting that "due to the very high labour factor involved in returnable bottles, we have had to, unfortunately, discontinue them."[40]

The issue of non-returnable soft drink containers faded from the public eye in the ensuing months. Behind the scenes, however, Pollution Probe began pushing the Ontario government to ban their sale in the province. On 20 May 1970, Minister of Energy and Resources Management George

Kerr announced a June meeting with soft drink container manufacturers "to discuss the whole question of non-returnable bottles and the litter problem." As Kerr explained, he would attempt to convince manufacturers to voluntarily stop using non-returnables; likewise, he hoped to persuade them to increase deposits paid for returnable soft drink containers from the current two cents to five cents on the grounds that it would encourage the public to bring their bottles back to stores.[41]

The Vickers and Benson Connection

Watching Pollution Probe's activities with interest was Terry O'Malley, vice president and creative director at the Vickers and Benson advertising agency. O'Malley grew up in St. Catharines, where local industry dumped untreated waste into the Welland Canal. Having previously taken this sort of behaviour for granted, he was heartened by Pollution Probe's efforts to clean up the environment and credited the group with raising his environmental consciousness: "I thought, 'You know, this is a chance for me to try and do something that I hadn't even thought of before.' I called them up and said anything I could do I would do *pro bono*."[42] Pollution Probe was at first skeptical of O'Malley's offer, considering that his agency's clients included major corporations such as Ford, McDonald's, and Gulf Oil. After a deputation met with O'Malley, however, it concluded that his offer was genuine. O'Malley developed a slogan for the organization – "Do it" – which highlighted Pollution Probe's belief that responsibility for addressing environmental issues lay with each member of the community. The slogan appeared immediately in all Pollution Probe documents and correspondence. As Peter Middleton notes, the Vickers and Benson connection "made an impact. It made us look professional,"[43] a significant achievement for an upstart organization with limited funding.

Pollution Probe now had a world-class advertising agency offering its services for free, but it still did not have the budget necessary for an ad campaign. This failed to deter Barrett, the inveterate optimist, who began a quest to obtain free print space from one of Toronto's prominent newspapers. Rob Mills recalls that this was one of many moments when Barrett's personality shone through:

> We headed down to get something from the *Globe and Mail* and they gave us a nasty "no." We went to the *Star* and they wouldn't even let us pass the front desk. *[laughs]* And Tony says, "Well hell, we're down here. Let's try

the *Telegram*." *[laughs]* ... I think I probably would have been the one that said, "Jesus, we've just been battered like hell, let's go back and think of another way to do it." And Tony's standing on the corner of ... King and Bay and says, "Well shit, its only five blocks to the *Telegram*. Let's try that."[44]

Barrett and Mills talked their way into a personal hearing with John Bassett, the *Telegram*'s owner-publisher, and convinced him to donate full-page advertising space to their fledgling organization. At first glance, Bassett and Pollution Probe appeared to be unlikely bedfellows. A prominent Tory, the businessman did not tend to sympathize with student activists. However, as Maggie Siggins explains in her biography of Bassett, the *Telegram* was on its last legs and struggling to find new niches within the Toronto newspaper market. It is likely that Bassett saw connecting with Pollution Probe as a way to appeal to an emerging audience, the environmentally conscious.[45] Pollution Probe's first full-page ad ran on 29 September 1969. Under the heading "How would you like a glass of Don River water?" the ad featured a black-and-white photo of a glass containing the river's sludge. With a description of the river's contents, an appeal to the public to raise their concerns with their political representatives, and an address to direct financial donations, the slickly produced ad was also the first to feature Pollution Probe's new "Do it" slogan. The advertisements continued on a regular basis until the *Telegram* closed shop in October 1971.[46]

The Don River Funeral

The Don River figured prominently in Pollution Probe's next major campaign. A dominant feature of the Toronto landscape, it served as a major waste sink for the rapidly industrializing city. With human sewage-induced bacteria levels as high as 61 million per 100 millilitres of test water – exponentially higher than the safe limit of 2,400 – the river running through the heart of Canada's largest city posed a health hazard to residents.[47] Although the general population was largely resigned to this, the members of Pollution Probe were not. Recent events in the United States suggested that the public's attitude towards the health of its waterways was beginning to change. A June fire on the Cuyahoga River in northeastern Ohio gained national attention, due largely to coverage by *Time*. Calls for a cleanup of the industrial sinkhole far exceeded those that followed the river's previous fires, which dated back to the nineteenth century. The

The Emergence of Pollution Probe 53

How would you like a glass of Don River water?

Isn't the Don River beautiful? Isn't it magnificent how it curls through the ravines of Toronto and flows beautifully into the Lake? Isn't it delightful how its banks have become the playground of children and families and other happy creatures?

It's nice to think what it could be. It's fun to think that the Don River could be pretty. That it could be useful to people. That it could be something other than a receptacle for the sewage that pours from the plants on its banks. That it could be something other than a stench that oozes into the air and makes you ashamed that it's there. And that this kind of thing could happen right here in the core of our city.

Don River water is full of everything raw and rotten you can think of. And Don River water doesn't stop in the Don River. It sort of gurgles and sludges its way with the wastes of all those progressive firms out into Lake Ontario. And the Lake goes to the St. Lawrence. And the St. Lawrence goes to the ocean. And after that, there's not much left is there?

Help us do something about things like the Don River. First write your mayor, or your Provincial Member of Parliament; or your Federal member; or even the Prime Minister. Tell them you'd like some of this stench cleaned up. If they don't believe it's there, or they give you some kind of song and dance, invite them over for a nice, cool glass of water. Don River kind.

We can use anything you'd like to contribute to help us work toward cleaning up our polluted air, water and land. If you would like to help us, we'll send you a button and a receipt for tax purposes, if you send us this coupon. Thanks.

Please Make Cheques Payable To:
UNIVERSITY OF TORONTO — POLLUTION PROBE

Name
Address
City

Do it. Pollution Probe at the University of Toronto.

FIGURE 5 Published as a full-page advertisement in the 29 September 1969 edition of the Toronto *Telegram*, "How would you like a glass of Don River water?" marked the beginning of a working relationship between Pollution Probe and Vickers and Benson. This ad was also the first to feature Pollution Probe's new "Do it" slogan. *Source:* Terry O'Malley fonds, RG 72, F2.65, Special Collections and Archives, Brock University.

summer also saw the maiden voyage of the *Sloop Clearwater*, a vessel designed to draw public attention to efforts to revive the Hudson River ecosystem. To draw attention to the Don as well as to the fragility of water ecosystems elsewhere, Pollution Probe decided to hold a mock funeral for the river.

Held on Sunday, 16 November, the Don River funeral began at 1:00 p.m. with a hundred-car procession, complete with a makeshift hearse, from

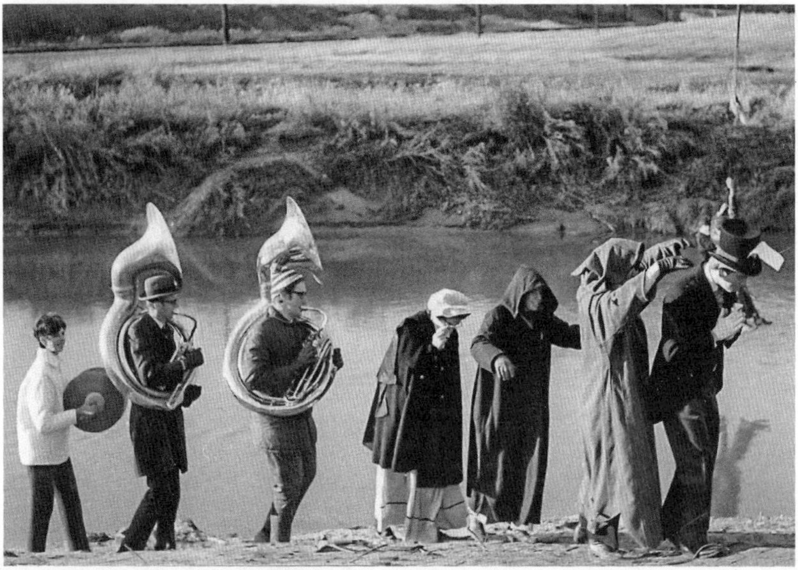

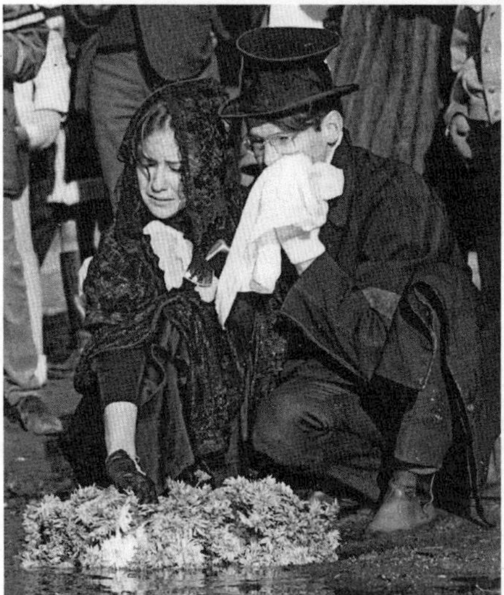

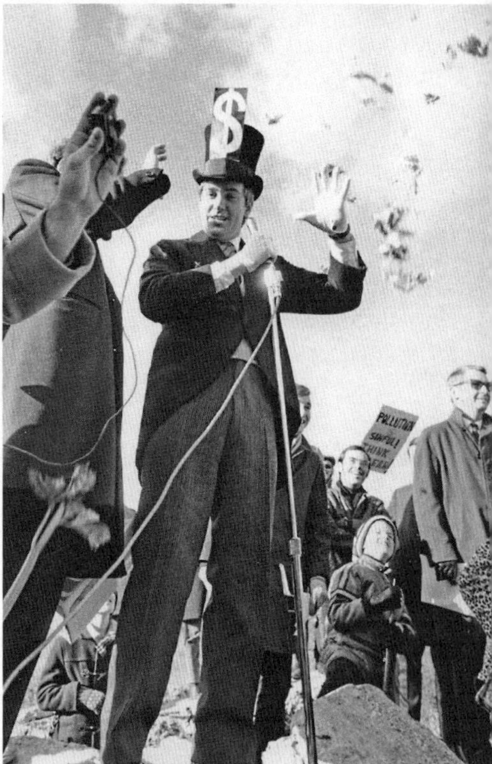
$
POLLUTION

the University of Toronto's Convocation Hall to the Prince Edward Viaduct. Represented by a black "coffin," the Don was then carried to the riverbank, accompanied by two sousaphone players performing a funeral dirge, and an estimated two hundred "mourners," many of whom were in period costume or holding placards bearing anti-pollution messages. Those in attendance listened to Martin Daly decry the state of contamination in the river, while Meredith Ware read descriptions of the Don's past grandeur from the diary of Elizabeth Simcoe, wife of Upper Canada's first lieutenant-governor. A forty-minute funeral ceremony was presided over by Rev. James Cunningham, the Hart House chaplain, who added some optimism to the event by predicting the river's eventual restoration. The funeral service's conclusion was punctuated by the appearance of Simon Greed, a wealthy industrialist portrayed by Tony Barrett. Wearing a tailcoat and a top hat, emblazoned with a large dollar sign, Barrett derided those in attendance, extolling the virtues of development and profit while minimizing the significance of pollution. As he wrapped up his speech, Barrett was pied by John Coombs, to applause from the crowd. In a final gesture, the funeral concluded with the placement of a wreath in the river.[48]

The theatrical nature of the Don River funeral is reminiscent of the guerrilla theatre common among contemporary activist groups. The art form emerged in 1965 from the San Francisco Mime Troupe. Using public areas as performance venues, the troupe aimed, in the words of founding director Ronnie Davis, to "teach, direct towards change, [and] be an example of change."[49] Utilizing humour, particularly satire, in order to impart the intended message, guerrilla theatre became most commonly associated

FACING PAGE, CLOCKWISE FROM TOP:

FIGURE 6   The Don River funeral, held on 16 November 1969, was a theatrical event that showcased Pollution Probe's ability to garner attention for environmental concerns while maintaining a light-hearted tone. In this image, members of the funeral procession are seen on the riverbank. Source: Personal collection, Tom Davey.

FIGURE 7   Simon Greed, a wealthy industrialist portrayed by Pollution Probe member Tony Barrett, derided those attending the Don River funeral while praising the virtues of development and profit. This photograph was taken moments before he was pied in the face by one of the mourners. Source: Toronto Telegram fonds, ASC17149, 1974-001/365, Clara Thomas Archives and Special Collections, York University.

FIGURE 8   The Don River funeral concluded with the placement of a wreath in the river. Pictured in this photograph is Meredith Ware (foreground, left), who read descriptions of the river's past grandeur from the diary of Elizabeth Simcoe. Source: Toronto Telegram fonds, ASC17148, 1974-001/365, Clara Thomas Archives and Special Collections, York University.

with the Youth International Party, or "Yippies," a US-based organization that gained considerable notoriety for its protests at the 1968 Democratic Party convention in Chicago. Pollution Probe's adoption of these tactics in the Don River funeral succeeded in garnering media attention, including front-page coverage in the *Globe and Mail*, national television coverage on *W5* and the *CTV National News*, and spots in the local CBC and CTV television news, the Toronto *Star*, and the *Telegram*. "I was on the cover of pretty much every newspaper across the country," recalls Ware. "I have an aunt in Vancouver and she phoned my dad and said 'Meredith's on the cover of the Vancouver *Sun!*'" *[laughs]*[50] All this coverage renewed interest in the state of the Don River, which became an area of interest for school field trips. It would also enter the political arena in August 1970 after a riverbank tour by federal Progressive Conservative leader Robert Stanfield, who remarked to the gathered media: "It's a bigger mess than I expected."[51]

The Citizens' Inquiry into Air Pollution

During this period, attention also turned towards Ontario Hydro. In September 1969, the Crown corporation announced plans to replace the smokestacks at Toronto's Richard L. Hearn Generating Station with a seven-hundred-foot "superstack." The coal-burning Hearn, which had a generating capacity of 1.2 million kilowatts, emitted 69,000 tons of sulphur dioxide in 1966, making it, in the words of *Telegram* reporter Mack Laing, "the worst single air polluter in the city."[52] The $9 million superstack, recommended in a 1968 report commissioned from Stone and Webster, was designed to alleviate the city's smog problem, particularly in the east end, by dispersing the effluent over a greater distance. As Ontario Hydro chairman George Gathercole explained before Toronto's Buildings and Development Committee on 22 September, "a higher stack reduces pollution by achieving greater dispersal or dilution." According to Gathercole, sulphur dioxide concentrations would be reduced by 90 percent in the city's downtown, and yet the effluent would not harm those living further downwind as it "is measurably weakened and changed by the combined influences of weather and dilution." He admitted that converting the station to natural gas would eliminate the sulphur dioxide pollution completely, but claimed that Ontario Hydro was unable to secure a steady supply of the fuel.[53]

Opposition to the superstack plan emerged immediately. Numerous letters to the editor appeared in the city's newspapers, suggesting that Ontario Hydro would better serve the public by investing in pollution-reducing technology. According to a letter published in the *Globe and Mail* by Thomas Beckett, chairman of the Hamilton and Region Conservation Authority: "It is most unfortunate that Ontario Hydro ... has now adopted the philosophy that the solution to pollution is dilution ... This will bring some relief to the individuals in the neighborhood of the plant. It will not reduce the total pollutants added to our atmosphere."[54] On 24 September, Dr. Ross Hall, chairman of the McMaster University Department of Biochemistry and one of the first to publicly condemn Ontario Hydro's plans, wrote Chant "to inquire whether you know if anyone plans to publicly raise the questions of human health and well-being related to the proposal."[55] Chant replied that this was "very timely because ... Pollution Probe is looking around for new projects."[56]

Pollution Probe kicked off its campaign in October with two advertisements in the *Telegram* attacking industrial air pollution. On 22 October, it issued a press release that raised numerous concerns with the proposal. After pointing out that sulphur dioxide kills the green plants necessary for producing oxygen, the press release emphasized that the "fact that the stack is higher does not get rid of the sulphur." Pollution Probe also sought clarification regarding Gathercole's statement that an adequate natural gas supply could not be secured, quoting Oakah Jones, president of the Consumers' Gas Company, as saying, "We can supply the gas if they'll tell us how long the plant is going to be in operation." The press release also challenged Gathercole's statement that a technology developed by the Monsanto Company enabling users to capture and resell sulphur dioxide was incapable of working at a generating station as large as Hearn. Dick Barnard, a Monsanto employee, was quoted as saying: "We gave Ontario Hydro a price on installing our system on one boiler, and haven't received a reply." The press release ended with a request for a copy of the 1968 Stone and Webster study. Pointing out that it was funded by $150,000 in taxpayers' funds, the press release noted: "The stack will not be private, and neither should the report be."[57]

Five days later, Pollution Probe member Paul Tomlinson, along with GASP's Tony O'Donohue and Drs. Ross Hall, Colin Locke, and J. Hodgins of McMaster University, attended a meeting with George Kerr and representatives from Ontario Hydro. That same day Pollution Probe ran its first advertisement taking direct aim at Ontario Hydro. Underneath the heading "The Ontario Hydro is getting ready to give it to you from great heights," it featured an ominous plume of black smoke emerging from a smokestack high above a crowd of onlookers. The ad highlighted the health and environmental problems associated with sulphur dioxide, the futility of simply spreading the Hearn Generating Station's effluent over a greater distance, and "the evident contradiction between the statements of Gathercole and Oakah Jones regarding natural gas supplies." The ad ended by encouraging citizens to write to Kerr to "register your feelings with him while you're still healthy enough to do something about it."[58] Following the meeting, O'Donohue announced plans for a public inquiry into the Ontario Hydro superstack, co-sponsored by GASP and Pollution Probe. Modelled after the dead ducks inquiry of July 1969, the organizers were promised full cooperation by Kerr and officials at Ontario Hydro.[59]

The Citizens' Inquiry into Air Pollution was held at New City Hall on 23-24 February 1970. The first day explored the general topic of air pollution in Toronto and featured headline-grabbing testimony from Dr. Joseph McKenna, a general surgeon at York-Finch Hospital, who explained that there was "irrefutable medical evidence that the air pollution of our atmosphere with extraneous material is responsible for more than 50 percent of all diseases seen in man." Furthermore, he blamed air pollution for a 700 percent increase in "respiratory cripples" in the city over the previous fifteen years.[60] The second day focused on the plan to build the superstack at the Hearn Generating Station. Gathercole presented a brief blaming those opposed to the project for "depriving people of a definite improvement in air quality in Toronto and surrounding areas."[61] He admitted that the superstack was only the beginning of improvements that needed to be made, but maintained that the sulphur dioxide would be diluted in the atmosphere, thereby eliminating a problem for the city's downtown and the surrounding areas. This testimony was sharply contradicted by Professor Benjamin Linsky of the University of West Virginia. Appearing via telephone, Linsky, a former pollution controller for Detroit and San Francisco, argued that without the installation of scrubbers, the superstack would merely serve as a "garden nozzle to spray" the sulphur dioxide further afield.[62] The inquiry's report recommended that the superstack be built on the condition that the Ontario government ban the use of fuels containing a sulphur content in excess of 1 percent. The report also recommended that the province alter its standards for sulphur dioxide to reflect the more stringent legislation in the United States, and to significantly increase research into air pollution.[63]

On 29 June, Gathercole announced plans to convert the Hearn Generating Station to natural gas by year's end. The move, which cost $4 million in renovations and led to an increase in rates, was made after Hydro signed a ten-year contract with the Consumers' Gas Company. The environmentalists' campaign was fundamental to this shift, as Gathercole informed the media that "anti-pollution measures are costly but our customers have indicated to us that they are prepared to pay for them."[64]

This was a high-profile victory, but the hard-fought battle against Ontario Hydro had broader implications for Pollution Probe. As its members explored the local issue, they realized that it was rooted in the growth ethos that dominated economic planning. As Brian Kelly explained: "At the time Ontario Hydro banked their whole business plan on a 7 percent annual growth in electricity consumption in Ontario. That caused us to say, 'Well, what about conservation? What about efficiency? What about alternative forms of generation?' And that got us into national energy policy issues."[65] This realization prompted further consideration of Canadian energy policy, which developed into one of Pollution Probe's central issues during the 1970s.

Breaking the Phosphate Impasse

Pollution Probe further solidified its national profile when it plunged into the ongoing debate concerning phosphate content in laundry detergents. During the first half of the 1960s, Canada and the United States dealt with the problem of "excessive foaming" in the Great Lakes, a problem that was resolved when industry switched to a biodegradable formula. Almost immediately, concern shifted to the massive algal blooms found on lakes, the product of cultural eutrophication. In December 1965, the International Joint Commission, an intergovernmental body charged with resolving issues in Canada-US boundary waters, urged the two governments to immediately reduce the amount of phosphate discharged into the waterways. These recommendations were non-binding, however, and little progress was made. A follow-up report, issued in October 1969, recommended the lowering of phosphate levels in detergents. This was fiercely opposed by the detergent industry, which countered that the best solution would be to improve sewage treatment facilities.[66]

Rather than waiting for industry and the various levels of government to come to an agreement, Pollution Probe took it upon itself to break the deadlock. A group of students led by Brian Kelly spent the Christmas 1969 holidays in Dr. Phil Jones's laboratory at the University of Toronto, analyzing the phosphate content of laundry detergents. The results were verified with industry and government scientists and released during a twelve-minute segment on CBC television's *Weekend* on 8 February. The list, read by Kelly and Peter Middleton, revealed a vast range in phosphate levels, from a high of 52.5 percent of total content in Amway Tri-zyme to a low of 10.5 percent in Wisk. When asked for recommendations on how consumers should proceed, Middleton urged them to use the low-phosphate options, noting that "the figures are out now – the consumer can make an intelligent choice."[67] Pollution Probe's television appearance had an immediate impact on the viewing public. By the end of March, over seven thousand requests for copies of the list poured into the organization's mailroom; the list was also reprinted in numerous magazines and newsletters. Consumer demand for this information proved so great that copies were prominently displayed in Loblaws, Dominion, and Steinberg grocery stores.[68]

On 9 February, the Ontario Department of Energy and Resources Management announced that it would reduce phosphate levels over five years. Pollution Probe believed that this phase-out was too slow, and in April presented Premier Robarts with a brief calling for a maximum phosphate content of 0.5 percent by January 1972.[69] Shortly thereafter the provincial and federal governments agreed to incorporate phosphate limits into the Canada Water Act, which would bring the legal limit down to 5 percent by the end of 1972. Although the federal government was already in the process of acting on the International Joint Commission's recommendations, and Ontario was considering following suit, Pollution Probe, argues historian Jennifer Read, "helped to concentrate public concern and kept the issue before the government while the parliamentary committee considered the legislation." Pollution Probe's greatest impact, however, was among consumers. Sales of high-phosphate detergents began to erode as low-phosphate options gained in popularity.[70] This was highlighted in the April 1970 edition of *Maclean's*, which documented the list's impact on West Hill, Ontario, housewife Rita Boston. Not only did Boston switch from Tri-zyme to a less harmful detergent – she also convinced her Amway saleslady to do likewise.[71]

Pollution Probe's Rising Profile

A telling sign of Pollution Probe's rising status can be gleaned from the pages of the *Globe and Mail*. In November 1969, Ontario Liberal leader

FIGURE 9 Pollution Probe's rapid ascent garnered the attention of elected officials from all levels of government. In this 1970 photograph, Robert Stanfield, leader of the Progressive Conservative Party of Canada, meets with members of Pollution Probe in the Ramsay Wright Zoological Laboratories. Pictured (left to right) are Dr. Ralph Brinkhurst, Terry Alden, Dave Lean, Tony Barrett, Robert Stanfield, Peter Middleton, and Monte Hummel. *Source:* Personal collection, Merle Chant.

Robert Nixon incorporated Pollution Probe into a speech to a gathering of the Ontario Student Liberals, stating that every campus across the province should have a branch of the organization. Pollution Probe quickly issued a press release emphasizing that it was politically nonpartisan,[72] but this would be just the first instance of politicians associating themselves with the fast-rising organization. On 3 March 1970, the newspaper ran a cover story about a speech by Prime Minister Pierre Trudeau at a Liberal Party fundraiser at the Royal York Hotel. The accompanying photograph showed Trudeau examining one of Pollution Probe's "Do it" buttons, which he had just been handed. Apparently he liked the button. As was reported: "After the dinner, the Prime Minister danced to the music of Ellis McClintock and the flashes of photographers. He wore a pink carnation and a Do it button."[73] Two months later, Opposition leader Robert Stanfield was in Toronto, drumming up support in a city that had not elected any Tories in the previous election, and was seen "sporting a pollution fighter's Do it button" while touring environmentally themed displays in Nathan Phillips Square.[74]

From the outset, one of Pollution Probe's central concerns was educating the general public on environmental issues. One of its programs involved sending speakers to schools throughout Metro Toronto. The presentations, which emphasized basic ecological concepts, the benefits of a healthy environment, and tips for living an environmentally friendly lifestyle, were seen as a vital component in empowering the public to make educated decisions. Beginning with just two speakers in June 1969, the program grew steadily as Pollution Probe's community profile increased. By March 1970, speakers' coordinator Stanley Zlotkin appealed to members to volunteer, noting that an influx of speakers would be necessary to accommodate bookings for the duration of the school year.[75]

Even as Pollution Probe proved adept at garnering the attention of media, government, and corporations, important changes were occurring behind the scenes. In the September 1969 edition of its newsletter, Tony Barrett, the organization's self-proclaimed "eco-financier," laid out its first budget, covering the next twelve months, which amounted to $54,750. Two months later, he released a revised budget covering the period from 1 October 1969 to 31 July 1970, for $79,600. These financial targets demonstrated an increase in ambition for the group, which, at the time of the second budget, had raised only $7,900, against $4,000 in total expenses. Most of Pollution Probe's early revenue came from the sale of pins and T-shirts bearing its logo, and from memberships. The latter's profit margin was minimal, however. In fact, memberships lost money in many cases, as they were originally sold for $2, less than the cost of mailouts. Effective August 1970, the membership fee was raised to $3 ($5 for non-students).[76]

The growth in Pollution Probe's planned expenditures coincided with the decision to make the group a full-time endeavour. Previously it had depended entirely on the energies of its student volunteers, including some who abandoned their academic obligations to focus on Pollution Probe's operations. Now, however, it was felt that a paid staff was necessary for continued growth. Four full-time coordinators were hired: Barrett, Brian Kelly, Paul Tomlinson, and Peter Middleton. According to Terry O'Malley, these four formed a sort of aggregate persona for Pollution Probe. "As one they were formidable," he recalls. "The intellectual [Tomlinson], the 'out front' guy in Tony [Barrett], the science guy in Brian [Kelly], and Peter [Middleton], the organizer."[77] Each was budgeted to earn $6,000 per year, although cash flow problems meant that they were typically paid just $250 per month.[78] Despite the minimal pay, the fact that they were being paid made them the first professional environmental activists in Toronto – and quite possibly Canada.

At twenty-five, Middleton was the oldest of the four. He was also the latest to join the group. A native of Etobicoke, "the suburban desert," he was the son of a bank manager. An avid Boy Scout in his younger days, and an experienced camper whose grandfather owned property along the Bruce Peninsula, he spent five summers working at Kilcoo Camp near Minden, Ontario, the first two leading the camp's nature lore program and the next three as a director. Valedictorian of his high school's graduating class, Middleton went on to study French at Victoria University in the University of Toronto, where he kept busy volunteering with the music club and heading the student council. "I was on the nerdy side," he recalls of his extracurricular activities. Upon graduating, he lived in Paris from 1966 to 1968 and witnessed first-hand the mass revolt in French society that led to the dissolution of parliament. The protesters' use of street theatre and the media made a lasting impact on him. In September 1969, he returned to the University of Toronto to pursue doctoral studies in French. The following month, after watching Larry Gosnell's third pollution special on the CBC, *Our Dying Waters*, he was moved to visit the Pollution Probe office and was greeted by the ever-present Tony Barrett. As Middleton recalls, "I made the mistake, so to speak, to ask 'Is there anything I can do to help?'" Subsisting on his salary as a don at the Victoria College residence, he quickly put his PhD studies on the back burner in order to work at Pollution Probe, where his extensive leadership skills were put to good use.[79]

Pollution Probe needed money in order to grow, and it set its sights on the Toronto business community. Its efforts to extend its base of support into the business community was both a natural progression and a deliberate policy. Initially, support came from the inner circle of its membership. Pollution Probe's first major corporate donation came in autumn 1969 from the North American Life Assurance Company. David Pretty, the vice president of finance, had been Rob Mills's scoutmaster in Lawrence Park. Mills recalls: "It was a natural fit because I knew the guy really well. I mean, it wasn't a big deal getting into his office … He was a fabulous scoutmaster. He took our group on canoe trips and [to] Temagami, and other places. He was a bit of a naturalist … so I think he just seemed like a totally logical person. I had no doubt he would support it."[80] Another important contact was Gage Love, president of W.J. Gage Limited and former chairman of the Toronto Board of Trade, whose son Peter and daughter-in-law Ann were among Pollution Probe's early members. According to Peter, "he certainly wasn't an environmentalist to start with, but he became very interested in it. And as it turns out, two of my brothers, and two of my sisters-in-law were also [eventually] staff at Pollution Probe. So he was pretty well surrounded at the dining room table."[81] These contacts were used to gain additional credibility for the group through its Board of Advisors, and to provide invaluable advice on such matters as fundraising. Some members of the business community took it on themselves to publicize Pollution Probe and its work among their peers. W.B. Harris, president of the investment banking firm Harris and Partners Ltd., was so taken with their "dynamic personalities and desire to undertake responsible research" that he organized a dinner at Hart House to introduce others to the group.[82] Likewise, Pollution Probe organized a two-day conference aimed at further incorporating this sector. As Mills wrote in a letter soliciting attendance: "We believe the business community has not been able to meet as a group to obtain a wide-ranging analysis of the cost and consequences of environmental pollution. Pollution Probe considers it essential to provide you with such an opportunity."[83] The 27 May session was designed "to enable you to come to a clear understanding of the ecological concepts involved in environmental contamination and man's place in the ecosystem," while the 3 June session, which featured the Honourable George Kerr among its speakers, aimed to clarify "Government positions on various pollution issues" and to "provide Business and Government with an opportunity to ascertain the responsibilities which lie ahead in the abatement of pollution."[84] Featuring an opening address by Bissell, whose support was sought "to assure the audience that our intent is honourable and that we are not just a radical student movement,"[85] the event was endorsed by such notables as Gage Love; Jim Gillies, head of the York University School of Business Administration; J. Bryan Vaughan, president of Vickers and Benson; F.S. Eaton, president of Eaton's of Canada; J.K. Macdonald, board chair at Confederation Life Association; and Raymond Moriyama, the world-renowned architect responsible for the Ontario Science Centre, the Japanese Canadian Cultural Centre in Toronto, and the Ottawa Civic Centre.[86]

Pollution Probe was even able to gain support from a company with which it had waged a public battle. In the aftermath of the phosphate campaign, Tony Barrett arranged for him and Brian Kelly to meet with the president of Procter and Gamble's Canadian operations. As Kelly recalls:

> We went in and told him about the phosphate thing and he certainly acknowledged the impact that it had on Procter and Gamble and so forth … It kind of appeared that we weren't going to get support from Procter

and Gamble. Towards the end of the meeting he reached into the desk drawer, pulled out an envelope, slid it across the table to Tony, and said, "Here, go and kick some more corporate ass." It was a cheque for $5,000, and that was big money in those days. But that was his expression. Corporately, he didn't want to admit that we really put the pressure on Procter and Gamble, but privately *[laughs]* he relished the fact that we were, in his terms, "Kicking corporate ass."[87]

Pollution Probe's relationship with the business community was ahead of its time. As Mark Dowie points out in *Losing Ground: American Environmentalism at the Close of the Twentieth Century*, ENGOs operating in the United States did not embrace corporate support until the 1980s.[88] Through the mid-1970s, for example, the venerable Sierra Club derived 70 percent of its revenue from membership dues, the sale of merchandise, and wilderness outings. By and large, partnering with business clashed with the notion of the New Left and counterculture that the root of environmental degradation was corporate greed; as such, financial support from corporations would be tainted money that provided legitimacy to an unworthy source. Pollution Probe's stance was largely influenced by Tony Barrett who, as a result of his upbringing, viewed the Toronto business establishment as potential allies rather than automatic enemies. More concerned with the ultimate outcome than the means of achieving it, Barrett was described by his peers as a pragmatic, middle-of-the-road reformer who stood in sharp contrast to the ideologues found elsewhere in the movement. It would be inaccurate to suggest that the environmental movement elsewhere was entirely devoid of moneyed interests – in the United States, many ENGOs received initial funding from private foundations, while the nationwide Earth Day celebrations in 1970 were partially funded by corporations[89] – but nothing matched the very open relationship between Pollution Probe and the Toronto business establishment. In short order, this became a model for ENGOs elsewhere in Canada, which sought Pollution Probe's assistance in establishing similar relationships.

By April 1970, Pollution Probe had grown into a major presence in the city. It had 1,500 members, four full-time coordinators, a secretary, and an office manager based out of the Ramsay Wright Zoological Laboratories. It was also a magnet for media attention, averaging an appearance once a week in the *Globe and Mail* and twice a week in the Toronto *Star*. It had made inroads in the local business community, and had found itself connected in the media with leading politicians. Despite the attention devoted to Pollution Probe's activities, however, it continued to emphasize the message that everybody had the ability to do good work on behalf of the environment. One could start by simply being conscientious about the amount of waste being generated, or by writing a letter to a politician asking for anti-pollution regulations. This message of personal agency also led Pollution Probe to encourage those living outside Toronto to develop their own independently operated affiliates. It was felt that having a network of environmental activist organizations in Ontario and across the country would help spread the heavy workload and increase recognition of the Pollution Probe brand. A guide for this process, "How to Form Your Own Pollution Probe," was created and sent to interested parties. Aside from providing the guide, which included advice on the best ways to start a group, establish its structure, draw public interest to its work, prioritize projects, and procure funding, those at Pollution Probe at the University of Toronto offered to send representatives to towns across the country to help organizers on the ground.[90]

The initiative proved to be a success, as affiliates soon sprang up across Canada. While the greatest concentration would be located in southern Ontario, where fifty affiliate groups were in place by the end of 1971, they could be found as far west as Regina and as far east as Halifax. These groups varied greatly. Many were relatively minor operations with a few keen environmentalists; others, such as Pollution Probe at Carleton University, had a paid staff and a broadly based agenda combining educational endeavours and political lobbying. Other affiliates would carve out specific niches for themselves. For example, the Peterborough group founded *Alternatives,* Canada's environmental studies journal, in 1971, while the Kitchener-Waterloo group contributed significantly to the advancement of recycling in the province.[91]

The sheer number of Pollution Probe affiliates that emerged across Canada demonstrates the national prominence the University of Toronto-based ENGO had attained. It also indicates that the country's environmentalists saw the use of the Pollution Probe name as a source of credibility within the broader community. This had a negative side, however. Whereas Pollution Probe at the University of Toronto worked hard to maintain its credibility, the emergence of affiliate groups meant that practically anybody could now speak on behalf of Pollution Probe. This failure to maintain a measure of quality control reveals a certain naïveté among those in the original group, whose desire to spread the environmental movement led them to overlook the potential harm that could result from uncontrolled growth.

The Emergence of Pollution Probe 67

FIGURE 10 The Summer Project '70 saw fourteen members of Pollution Probe, in teams of two, travel throughout cottage country discussing water-related environmental issues. Over the course of the summer, they spoke to an estimated 25,000 campers and cottagers. Pictured is the team of Brian and Ruth Kelly. *Source:* Toronto *Telegram* fonds, ASC17108, 1974-002/160, Clara Thomas Archives and Special Collections, York University.

Pollution Probe's attention also turned to cottage country. The organization saw the "pollution of our inland waterways and lakes and the ruin of campsites and parklands as a life and death question for both camps and the resort areas in general."[92] Its Summer Project '70 was designed to educate campers and cottagers on environmental issues concerning the water and to motivate them to find solutions. Pollution Probe was willing to devote its summer program to cottage country because, as previously noted, summer camping had been a formative experience for many of its members. It also reasoned that working with the established associations would enable it to reach affluent Ontarians, an important consideration for the ambitious ENGO.[93]

In many respects, Summer Project '70 was Pollution Probe's most ambitious undertaking yet. Seven two-person teams were hired and assigned regions. Funds for the project, which cost $45,000, came from a variety of public and private sources, including the Ontario Water Resources Commission, the federal Department of Energy, Mines and Resources

(EMR), the National Capital Commission, John Labatt Ltd., Loblaw Groceterias, the White Owl Foundation, the *Globe and Mail,* the Toronto *Star,* and Coles.[94] The project marked the initial collaboration with an affiliate, in this case, Pollution Probe Ottawa, which provided one of the teams. It also required the coordination of schedules with the cottage associations and camp owners, as well as with local newspapers and radio stations, whose help was enlisted to publicize the coming visits.

After spending May and June preparing, the seven teams hit the road in July. Over the next two months, they spoke to an estimated 25,000 people. The major environmental hazard was found to be inadequate sewage treatment, particularly among individual cottagers, and a lack of nutrient-removing facilities in the local community sewage plants. While there was discussion of infrastructure developments that could alleviate this problem, the teams also discovered a significant problem of overdevelopment on the lands in question. Noting that lots as small as seventy-five feet across were being sold, the teams highlighted the need for individual municipalities to pass retroactive by-laws concerning minimum lot sizes. The teams also discovered that stores and laundromats in cottage country still favoured high-phosphate detergents, which had been at the centre of the recent eutrophication problem in the Great Lakes. Besides educating local residents on the necessity of addressing these problems, the Pollution Probe team taught cottagers how to test their own water and encouraged them to establish a system of self-policing. On completion of the project in late August, a report of the findings was assembled and distributed to cottage associations, government, and the project sponsors.[95]

POPULATION CONTROL AND THE ENVIRONMENTAL MOVEMENT

One of the more contentious aspects of first wave environmentalism that Pollution Probe struggled with was the neo-Malthusian argument that the ever-growing human population was primarily responsible for the planet's environmental degradation. This concept traced its roots back to the work of Thomas Robert Malthus, a British scholar best known for his 1798 publication *An Essay on the Principle of Population.* According to Malthus, population increases in geometric progression, whereas subsistence increases arithmetically. Left unchecked, he argued, population would inevitably outstrip subsistence, leading to calamity. The concept was revived in the postwar period by Fairfield Osborn's *Our Plundered Planet* and William Vogt's *Road to Survival,* two 1948 environmental treatises.

Historian Samuel Hays has linked the publication of these books to a postwar attitudinal shift "from optimism to a guarded pessimism" ... "both of them [are] infused with Malthusian pessimism, both emphasizing the enormous problem of population growth and the world's limited food supply. Both warned that technology was not enough; resources were not unlimited; the pressure of population itself must be reduced."[96] This concern would reach its apogee with the 1968 publication of Paul Ehrlich's *The Population Bomb*. An entomologist at Stanford University, Ehrlich successfully brought the message of population control to the mainstream, as evidenced by the millions of copies sold and his many guest appearances on the *Tonight Show Starring Johnny Carson* between 1970 and 1972. Ehrlich's environmental jeremiad opened with the declaration that

> The battle to feed all of humanity is over. In the 1970's the world will undergo famines – hundreds of millions of people are going to starve to death in spite of any crash programs embarked upon now. At this late date nothing can prevent a substantial increase in the world death rate, although many lives could be saved through dramatic programs to "stretch" the carrying capacity of the earth by increasing food production. But these programs will only provide a stay of execution unless they are accompanied by determined and successful efforts at population control. Population control is the conscious regulation of the numbers of human beings to meet the needs, not just of individual families, but of society as a whole.[97]

The Population Bomb, which attributed all environmental problems to overpopulation, gained prominence alongside *Silent Spring* as a must-read for those concerned with the state of the planet.

Capitalizing on the attention captured by his book, Ehrlich and his colleagues launched Zero Population Growth, a group dedicated to "press[ing] for legislation to implement far-reaching birth control programs" and the "allocation of funds for more research into population problems and research for better methods of contraception."[98] By 1970 the organization, which urged parents to "Stop at Two," had 380 chapters and 33,000 members across the United States. More important, by the time of the first Earth Day, many of the leading environmental groups in the United States had adopted, or were considering adopting, population control as an important part of their platforms.[99]

As noted in Chapter 1, the University of Toronto played host to an international teach-in on population issues in October 1968. The event, headed by Drs. Henry Regier and J. Bruce Falls of the Department of

Zoology, featured a number of prominent participants, including Donald S. Macdonald, the president of the Privy Council; the Reverend Dr. Frank P. Fiddler, the past president of the National Council of Churches in Canada and president of the Family Planning Federation of Canada; Father Gregory Baum, a St. Michael's College-based theologian who served as a consultant to the Second Vatican Council; and George Cadbury, the former president of the New Democratic Party of Ontario and a one-time executive director of the International Planned Parenthood Federation. Cadbury and his wife, Barbara, wealthy British immigrants who were prominent in the local and international birth control movement, were largely responsible for bankrolling the event.[100] Dr. Chris Plowright, an Englishman who joined the University of Toronto's Department of Zoology shortly before this event, had a concern for population issues dating back to 1960, when he read *Adam's Brood: Hopes and Fears of a Biologist,* by prominent British eugenicist Colin Bertram. Plowright recalls: "That book was a shock because it had never occurred to me that human numbers were a threat to the planet." He was enthused by the response to the teach-in, which drew over three thousand participants and significant media coverage. As he notes wryly, "Some of us, in our ignorance and naïveté, were even encouraged to think that maybe this was going to make a difference."[101] In March 1970, he headed the launch of Toronto Zero Population Growth (TZPG), an independent affiliate of Ehrlich's organization.

Like Pollution Probe before it, TZPG received the support of Dr. Donald Chant, who provided it with office space on campus. In fact, there was considerable overlap between the two groups. While Pollution Probe failed to undertake any sustained campaigns on neo-Malthusian grounds (with the notable exception of the Energy and Resources Project, discussed in Chapter 3), many of its members were firm believers in the link between population growth and environmental degradation. Pollution Probe greeted the launch of TZPG with open arms, noting in the *Probe Newsletter* that "the issues of pollution and population growth are inseparable. Pollution Probe welcomes the birth of ZPG-Toronto and has decided to hand over its work on the population problem to its new little sister."[102]

Those promoting the population control message had a difficult message to sell. As Donald Worster explains: "Here the environmentalists confronted deeply seated attitudes among traditional economists, business leaders, politicians, and the public about the virtues of economic growth, attitudes underlying the modern economic system and indeed the whole materialistic ethos of modern culture."[103] More importantly, support for population control challenged common moral and ethical codes pertaining to human sexuality and reproductive rights, and was fiercely opposed by groups such as the Catholic Church. As Ralph Brinkhurst, himself a supporter of TZPG's fundamental message, notes: "The whole idea of imposing population limits on people is a whole lot harder to sell than the idea of cleaning up the environment, which could be, and too often is, hitched to an idea that it is about human health."[104] A major challenge facing population control advocates in Canada was that the country's low population density, coupled with its wealth of natural resources, rendered their claims of imminent population-induced apocalypse difficult to fathom. Likewise, Canada's birth rate was already declining steadily. As Premier John Robarts wrote to a TZPG member in May 1970: "Where overpopulation may become a problem on a world basis some time in the future, it is certainly not the case in Canada nor even here in Ontario ... As a matter of fact, the birth rate in Ontario has been dropping over the last few years and will likely continue to do so."[105] Against this socio-cultural setting, TZPG failed to take root in Canada, peaking in 1971 with a total of eight independent chapters and approximately five hundred members, most of them in Ontario.[106] According to Plowright, TZPG felt isolated from the rest of the environmental movement: "Pollution Probe, in all its public statements, they never would say anything about population growth, and that was generally the thing in those days. Us Zero Population Growth people were [depicted as] nuts on the left fringe and the middle of the environmental movement preferred not to get into it, for obvious and very good reasons."[107] Although this was not entirely accurate – members of Pollution Probe had gone on the public record advocating population control as an environmental necessity – minutes from a 17 February 1971 Pollution Probe meeting reveal that it rejected the idea of absorbing TZPG because it was "felt that we would be labled [sic] as ZPG and this would hamper our effectiveness."[108] Likewise, there was considerable hesitation within Pollution Probe about taking on any population projects, for fear that the organization would be confused with TZPG.

The idea of absorbing TZPG was on Pollution Probe's agenda because the population group began to fall apart almost as soon as it was formed. Led by Plowright and Dr. Dennis Power, an evolutionary biologist at the Royal Ontario Museum, neither was able to devote the full-time energy necessary to properly launch such a project. Furthermore, as Power notes, "I was naïve enough in those days to not even think about having to incorporate as a nonprofit organization. Anything smacking of 'business' on top of academic work may have taken [away] some of our missionary zeal."[109] This absence of business acumen was aggravated by Plowright's personal difficulty addressing the subject. As he explained in an interview, "I found nothing more depressing than working on population control [and] population problems. It's just the most awful, horrible, miserable, depressing subject possible to imagine. I eventually sort of retreated and gave up because I couldn't stand the depressive pressure of it."[110] TZPG continued to operate throughout much of the 1970s, but amounted to little more than an information service operated by schoolteacher Janice Palmer.

GASP's Last Gasp

On 10 April 1970, GASP held its first "annual" meeting at City Hall. Just twelve people turned out. O'Donohue informed those in attendance that GASP's finances were in shambles. He had hoped to raise $145,000 but GASP had only $178.70 in its coffers. He noted that very few of the group's 450 members had bothered to pay their $2 yearly dues. He had also been rejected by all but one of the union locals he had approached for support. Other organizations exhibited, in O'Donohue's words, "tremendous support" for GASP's work but stopped short of donating to the cause. O'Donohue rejected the idea of asking industry for funding, arguing that it could place the group in a compromising position.[111]

GASP's final undertaking, in conjunction with Pollution Probe, was a "Leave the Car at Home Week," to be held from 12 to 19 July 1970. O'Donohue explained at the initiative's public announcement: "We want people to walk to work, take public transportation or form car pools so we can measure the effect fewer cars in downtown Toronto has on air pollution." He hoped for City Hall's cooperation, particularly in closing a number of streets in the downtown core. Fellow GASP member James Karfilis added that closing the downtown area would enable scientists to determine whether there was a significant decline in carbon monoxide and nitrous oxide, which could have a long-term impact on city planning.[112]

Unlike previous GASP-Pollution Probe collaborations, this one saw GASP taking the lead role. Unfortunately, its efforts were for naught. The city's Public Works Committee initially seemed open to a partial closure of Bay Street, but it was noted that this would leave the city open to being sued, under the Municipal Act, for financial damages suffered by local businesses; the committee also rejected the idea of having those affected sign waivers relieving the city of liability.[113] In the end, the event was formally delayed. A joint GASP-Pollution Probe press release explained that the decision was the result of three factors. The groups claimed to have underestimated the popularity of the event and, given their lack of resources and the short notice, felt that a delay would be vital to its success. They also cited the desire to reverse the Public Works Committee's decision to keep Bay Street open to automobile traffic. Finally, they noted: "We want to provide more than a public relations campaign to encourage people to try other transit. While we are a minority committed to seeing cleaner air in Toronto and we realize our serious air problem, we know that most people will not take public transit unless there are lower fares (or none at all) and increased convenience." In line with this, they announced a plan to co-sponsor "a citizens [sic] inquiry into the pollution controls and fuels available for cars and to survey transit systems in Toronto and in other parts of the world."[114] Shortly thereafter, however, GASP ceased operations for the second and last time, as O'Donohue focused on his 1972 mayoral bid.

Conclusion

By the summer of 1970, Pollution Probe was firmly established as a pillar of the burgeoning environmental community, with a paid core staff, a rising profile, and a series of affiliate groups across Canada. Meanwhile, GASP was defunct. This may seem surprising at first blush, given that GASP was launched just weeks after the highly controversial broadcast of *The Air of Death*, and it appeared to have the backing of affluent professionals.

While it is true that GASP was established by a group of prominent Torontonians, it remained a part-time pursuit. Its high-profile public launch with three hundred in attendance did not translate into an active membership, and it relied primarily on the work of a small group of individuals throughout its brief history. The initial public meeting was also held just over a month after the broadcast of *The Air of Death*, and it is possible that many spectators were drawn by Stanley Burke's presence as the meeting's moderator, given his popularity and the media controversy surrounding the television program. With its key members preoccupied with full-time jobs and family responsibilities, the organization did not have the opportunity to pursue government grants or other forms of funding necessary to hire staff and fund projects. And although GASP did acquire a full-time executive director in the person of Tony O'Donohue in January 1969, this failed to reverse its fortunes as it tied the organization to the alderman's political ambitions.

Pollution Probe, on the other hand, benefited from the support it received from the Department of Zoology at the University of Toronto. Being provided with office space, telephones, and a forwarding address enabled Pollution Probe to continue its operations without worrying about burdensome overhead costs, and its association with the university accorded it credibility. All of this would have been meaningless, however, without the organization's dedicated volunteers and membership. The small core of volunteers who provided the group with direction could call on a paid membership that reached 1,500 in April 1970, enabling them to orchestrate newsworthy events such as the Don River funeral. These factors helped Pollution Probe's fundraising efforts, which in turn enabled it to hire staff and become a full-time operation.

Seen from the standpoint of resource mobilization theory, the reason for Pollution Probe's success vis-à-vis GASP is even more clear-cut. Originated by sociologists John D. McCarthy and Mayer N. Zald, this theory argues that formal social movement organizations, including ENGOs, operate in a manner akin to firms insofar as they aspire to accumulate resources, employ staff, and sell their work to potential financial contributors. Just as retailers compete to attract business from a limited clientele, ENGOs must compete with one another for funding. In order to alleviate competition, ENGOs often specialize, but sometimes they go head-to-head for funding.[115] Despite their common goal of combating pollution, the two organizations ultimately competed with each other for funding. Pollution Probe established itself as a media darling with its high-profile activities, but it was easy to overlook GASP, which did little beyond the two public inquiries it co-sponsored with Pollution Probe. Unable to differentiate itself from its more youthful counterpart, GASP was doomed to compete for funding in a head-to-head competition that it could not win.

3
Building an Environmental Community

O N 14 OCTOBER 1971, a crowd of Pollution Probers, their supporters, and interested members of the media convened on Parliament Hill. Over the course of the preceding seven days, the ENGO's Resources Recycling Caravan had followed a serpentine route throughout Ontario, gathering metals, paper, and other recyclables. In the process, Pollution Probe filled one tractor trailer and five additional trucks. Now they awaited the appointed time to publicly deed the collected materials to Jack Davis, Canada's environment minister, to serve as seed money for a national recycling initiative.

This event highlights two developing themes for Pollution Probe. Between autumn 1970 and the end of 1971, Pollution Probe began expanding its focus beyond its Toronto stronghold. Recognizing that it was ideally suited to provide leadership to the far-flung and fledgling ENGOs nationwide, it began constructing the first stages of a Canadian environmental community. This involved organizing the Canadian Association on the Human Environment (CAHE), the first body to unite environmental activists across Canada, as well as putting on workshops to teach fellow upstarts how to orchestrate successful action and fundraising campaigns.

The Resources Recycling Caravan also highlights the broadening of Pollution Probe's parameters. Up to this point, the ENGO had earned its reputation addressing end-of-pipe pollution issues. As rapidly became apparent, however, pollution was only one aspect of environmental degradation. Recognition of the need to address the underlying environmental problems led to the autumn 1970 initiation of the Energy and Resources

Project, which cited a link between Canada's energy sector and the consumer-driven growth ethos that imperilled modern society. Even as the ENGO continued to address matters of air, land, and water contamination, its broadened perspective resulted in its rebranding. "Very quickly it wasn't Pollution Probe, it was Probe," explains Peter Middleton, "because pollution was just one angle."[1] The full name was retained for legal reasons, but the ensuing publications and promotional materials featured the shortened version.

This period also saw significant developments in the local environmental infrastructure. Pollution Probe continued to foster the growth of an assortment of complementary ENGOs, most notably the Canadian Environmental Law Association (CELA), which was formed to provide local environmentalists with a legal arm. Some of the most significant developments occurred within Pollution Probe itself. Continued growth – September saw its staff grow from four to sixteen – rendered the consensus-based model of operations increasingly inefficient. Tensions flared as the decision-making process bogged down. With key staffers threatening to resign, the position of executive director was created in the summer of 1971 and became a key step in Pollution Probe's professionalization.

Money Matters

Pollution Probe's growth from a base staff of four to sixteen in September 1970 was made possible by a concomitant increase in funding. That month saw the release of Tony Barrett's latest budget, which called for $291,100 in expenditures over the coming year – an increase of nearly sixfold over the rather modest budget introduced a year earlier. A new Board of Advisors was created to aid in fundraising. The initial board placed heavy emphasis on scholars, a logical move for a young ENGO still in the process of establishing its credibility; the new board had just a single holdover, Dr. Chant, alongside five prominent industry leaders: R.D. Brown, a partner at Price Waterhouse; J.H. Davie, vice president and director of Dominion Securities Corporation; C. Malim Harding, chairman of the board at Harding Carpets; D.W. Pretty, vice president at North American Life Assurance Company; and David Purdy, Vice-President Finance at Imperial Life Assurance Company of Canada.[2]

The September 1970 budget and the coinciding shakeup of the Board of Advisors signalled a new, aggressive approach to fundraising, but Barrett recognized that Pollution Probe was fighting an uphill battle in securing funds. He wrote in the *Probe Newsletter:*

> Donors usually have policies of giving to causes or charitable organizations within defined categories they choose – we don't fit into anyone's category so new ground must be broken with most prospective donors ... [Furthermore] the general economic climate being what it is, donations budgets are facing cuts. The result is that the established charities like hospitals, schools, United Appeal are given priorities [sic], emphasizing that the broad spectrum of environmental ills are to a considerable degree a compounding or basic factor in other more traditional problems and that donations should be going more to root problems.[3]

Barrett's caution was warranted. Pollution Probe's 1971 year-end budget reveals that the ENGO raised $184,805 over the preceding twelve months. Although this represented a significant increase in the organization's fundraising capacity, it also fell short of the goal for the year.[4] No breakdown of the sources of revenue is provided, but a review of the page titled "Major Corporate Donors – 1971" is telling. Three donors were foundations and four were government bodies; the remaining seventy were corporations. This reliance on government, private foundations, and corporations became typical of Pollution Probe's revenue stream over the next decade.

Despite Pollution Probe's professionalization, as evidenced by the expansion of its paid staff and the creation of the Pollution Probe Foundation in June 1971, which gave it charitable status, little changed in the way it operated. There was still a degree of creative anarchy, as individual members were encouraged to undertake whatever projects struck their fancy. This freedom was exemplified by staff member Terry Alden, a Massachusetts Institute of Technology graduate who pursued projects ranging from a Donner Canadian Foundation-funded study of noise pollution in the city to an exploration of the effects of radiation pollution from the Pickering Generating Station.[5]

Book Release and Survival Day

After months of careful preparation, autumn 1970 witnessed the public launch of two projects. The first was the release of *Pollution Probe,* a 209-page guide to living an environmentally friendly life. Written by the organization's membership and edited by Donald Chant, the book provided an overview of leading environmental problems and practical solutions. Released by the New Press, an upstart publisher headed by James Bacque, the former chief editor at Macmillan Company of Canada and founder of GASP, the book sold out its first printing of 6,200 copies within a month of its 30 September launch. This might have provided a welcome source of revenue, but concern that the intended $3.50 cover price might limit its readership led Pollution Probe to subsidize the first batch, dropping the price to a more "affordable" $2.50. A second printing quickly followed.[6]

As *Pollution Probe* was being released to the public, the organization was finalizing a full slate of activities to help celebrate Survival Day. Slated for 14 October 1970, Survival Day emerged as a Canadian equivalent of Earth Day. Although the first Earth Day celebration on 22 April 1970 involved twenty million Americans, it was a relatively minor event in Toronto, highlighted by an all-day vigil at Queen's Park that drew a peak crowd of one hundred, including members of the provincial Liberals handing out packets of phosphate-free laundry detergent in Nathan Phillips Square.[7] Pollution Probe chose to skip the first Earth Day entirely. As Brian Kelly explained to a *Globe and Mail* reporter: "As for Earth Day, let the United States do that and it's great. But it's the wrong time for us, right in the middle of exams, and we have to rely on students. We have Oct. 14 as a tentative date to do our own thing in Canada – major speeches, tours of pollution highlights, and so on."[8]

Pollution Probe organized six days of activities in and around Toronto leading up to Survival Day. The appropriately named Survival Week, beginning 7 October, featured seminars, public lectures and debates, tours of conservation areas and a local sewage facility, a bicycle parade from Lawrence and Yonge to City Hall, and performances by Pro-Seed, an ecologically themed theatre group. Survival Day itself featured the burying of a time capsule at the site of the planned Humanities Research Library at the University of Toronto. Lowered into the ground by Chant, the capsule contained vials of DDT and water from the Don River, a recording of noise pollution in the city, various newspaper clippings concerning environmental degradation, and a bronze plaque with an apocalyptic message: "In the hope that this time capsule will be found by a civilization wiser than our own, we have buried here a record of man's folly on the planet he has outgrown."[9] There was also a gathering of 250 high school students at Convocation Hall to hear Stanley Burke and Dr. Chris Plowright discuss methods of addressing environmental problems, and a "general pollution debate" hosted by Drs. Claude Bissell and Donald Chant, and

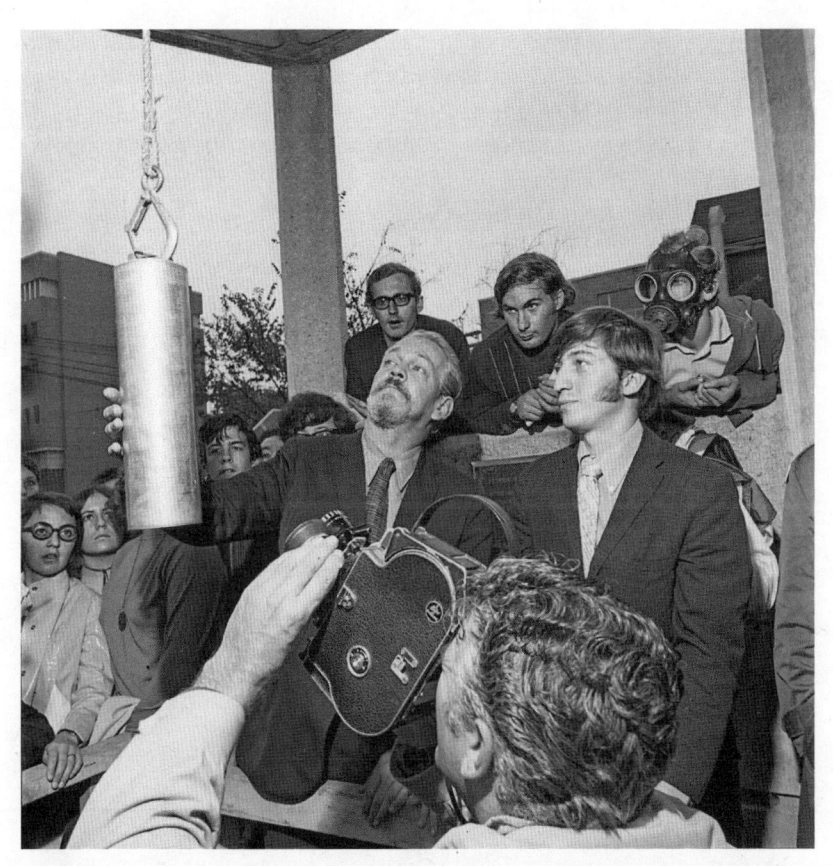

FIGURE 11   As part of its Survival Day activities, Pollution Probe buried a time capsule featuring "a record of man's folly" outside the present-day John P. Robarts Research Library, October 1970. Holding the time capsule is Dr. Donald Chant. *Source:* Robert Lansdale fonds, B1998-0033 [701283-22], University of Toronto Archives.

featuring Donald Collins, chairman of the Ontario Water Resources Commission, Liberal MPP Murray Gaunt, and NDP MPP Fred Young.[10]

The turnout for Survival Week events in Toronto was discouraging. The bicycle parade was considered a success – despite heavy rain, approximately two hundred turned out, including Liberal power couple Stephen and Adrienne Clarkson on a tandem bicycle – but most events were sparsely attended. Ontario's seventeen community colleges spent $10,000 advertising on-campus seminars, which also had disappointingly low turnout. Furthermore, despite national ambitions for Survival Day, it did not seem to make much of an impact in any other major cities, except perhaps in Ottawa, where the Board of Education authorized schools to devote their afternoon classes on 14 October to anti-pollution activities and education. Survival Day's lack of success can be at least partially attributed to its timing – it had the misfortune of occurring in the midst of the then-unfolding October Crisis that saw the Canadian military dispatched to the streets of Montreal following a series of political kidnappings by the Front de libération du Québec – which left little room for the nascent event in national headlines. Initial plans to revive the event in 1971 were abandoned.[11]

Birth of the Energy and Resources Project

In 1964, the governments of Lester Pearson and Lyndon Johnson commissioned a study on bilateral relations between their countries. Among the points made in the highly influential Merchant-Heeney report that resulted was "the economic advantages to both countries of disregarding the boundary for energy purposes."[12] This idea of a continental energy pact gained momentum in 1969 when J.J. Greene, the Canadian minister of energy, mines, and resources went on public record in favour of the concept, proclaiming that "people will benefit, and both countries will benefit, irrespective of where the imaginary border goes."[13] That same year, the Nixon administration appointed a Cabinet Task Force on Oil Import Control, chaired by George P. Schultz, to examine how the United States could double its consumption over the next thirty years despite the fact that domestic energy production was levelling off. The task force report, released in February 1970, advocated the establishment of a continental energy pact with Canada.[14] Negotiations between the two countries were scheduled to begin in November 1970.

Concern over the proposed pact gave rise to Pollution Probe's initial work on energy and resource issues. This was an unintended offshoot of its earlier battle against Ontario Hydro's plans to build a seven-hundred-foot "superstack" at the Richard L. Hearn Generating Station. While the initial concern was that Ontario Hydro would merely disperse the station's sulphur dioxide effluent over a greater area instead of reducing the pollution, attention soon shifted to the growth ethos guiding the corporation's business plan. Pollution Probe's Energy and Resources Project, first described in the October 1970 edition of the *Probe Newsletter*, was the undertaking of Brian Kelly and Geoff Mains. Both Kelly and Mains came from a scientific background, a rarity among Pollution Probe's active membership. Kelly held a Bachelor of Science degree while Mains was pursuing a PhD in biochemistry at the University of Toronto. Although issues pertaining to energy and resources did not appear to be of immediate concern to an organization renowned for battling pollution, closer examination revealed that they were integral. Echoing the neo-Malthusian concerns raised by Paul Ehrlich in *The Population Bomb,* and foreshadowing the message of the Club of Rome's 1972 *Limits to Growth,* which used complex computer modelling to demonstrate the deleterious consequences of population growth and the strain on finite natural resources, the Energy and Resources Project placed the growth ethos at the centre of all environmental problems. As Kelly and Mains explained in Pollution Probe's October 1970 newsletter:

> In striving for a quality environment, uncontrolled economic and population growth is the basic problem which we must all attack, for the growth ethos of our modern society is undoubtedly the major underlying cause of most environmental problems. We should all be devoting more time and effort towards attacking these root causes, for without progress on this front all other forms of anti-pollution work will be for naught.[15]

The Energy and Resources Project was openly critical of the consumer-driven lifestyle of North Americans: "On a global level, if North America can demonstrate self-control and restraint in growth and consumption there is little reason why other countries could not follow."[16] Given the United States' large population and international influence, it became the central focus of much of the project's activities. Home to just 6 percent of the global population, the United States consumed roughly one-half of the world's available resources.

Since a continental energy pact would fuel American growth with Canadian energy and resources, the Pollution Probe project opposed it. Using the November 1970 negotiations as a launch pad for attacking the root problem of North American overconsumption, Kelly and Mains initiated what they characterized as the first phase of "a massive long-term project" in October. A major problem, they reasoned, was the fact that the United States had a clear-cut aim of improving its access to Canadian energy and resources, whereas Canada did not have a firm policy in place pertaining to its own energy and resources. They argued that Canada needed to formulate such a policy, and that it should include using "its resources as a lever to force the United States into specific programs of population control, restraint in economic growth, and recycling." Since the United States would naturally demand that Canada adopt similar programs, this was viewed as a win-win situation. Consequently, in advance of the November talks between the two countries, Kelly and Mains announced their intention to sell the following three points to the Canadian government:

1 that it should make no commitment towards a Continental Energy Pact or resource sales at the November meetings;
2 that it should make no agreements until a Canadian Energy and Resources Policy is formulated; and
3 that it should seriously consider using Canadian resources as a lever against continued American growth and consumption.[17]

Kelly and Mains sought to rally popular support for their initiative. Letters were sent to one hundred environmental groups across Canada, and to another 340 prominent American groups and environmentalists. Recipients were encouraged to write letters in support of the three-point agenda to J.J. Greene, Prime Minister Pierre Trudeau, and their federal representatives, with carbon copies sent to Pollution Probe for tracking purposes. The hoped-for groundswell of support failed to materialize, however: by 30 November, Pollution Probe had received just twenty-seven positive responses from the Canadians, forty-three from American groups, and just two from individuals.[18] On 9 November, it took the campaign public with a full-page advertisement in the Toronto *Telegram*. Under the heading "Now that we've nursed the hungry Monster through its gas pains, what will we feed it next?" the ad featured a giant wearing a top hat emblazoned with the Stars and Stripes sitting outside a castle flying the Canadian flag. As the giant indulges in a hookah labelled "Canadian Natural Gas Resources," a group of people carry a water pipe across the castle's drawbridge. The accompanying text explained that the talks in November were designed to increase US access to Canadian energy and resources, and that Canada's lack of an energy and resources policy made it ripe for exploitation by the Americans. The proposal to use Canadian energy and resources as a lever was highlighted and signature-ready coupons were provided, expressing concern to the Honourable J.J. Greene.[19] On 16 November, Brian Kelly took the message to CBC television's long-running *Viewpoint* program, delivering a five-minute presentation urging the general public to support the three-point plan. With Stanley Gershman of Toronto Zero Population Growth (TZPG), he also sent a starkly worded letter to the editor of the Toronto *Star*. Published in the 17 November edition, the letter described the United States as "an insatiable monster" and argued that the

> impending result of this glutinous [sic] consumption will shortly be depletion of resources vital to our civilized way of life, irreversible degradation of our environment, extreme and permanent deprivation of a decent standard of living for the majority of an exploding world population, and the certain continuance and spread of warfare as the deprived people of the world become increasingly dissatisfied and aggressive in their demands for a share of the world resource pie.

The only way to avoid this perilous outcome, Kelly and Gershman wrote, would be for Canada to adopt Pollution Probe's three-point plan.[20]

The message was also delivered directly to Parliament. Mitchell Sharp, the secretary of state for external affairs, was predictably defensive over the government's record, noting that despite recently approving an increase in authorized gas exports to the United States, he wanted "to ensure that there were adequate proven reserves to meet Canadian needs over and above those committed to the export market." He also dismissed the proposed lever approach: "It is not clear to me ... that the approval of the export of a specified volume of natural gas to the United States represented an appropriate opportunity to deal with matters such as a population control programme and an end to uncontrolled growth."[21] Undeterred, Pollution Probe sought and received a meeting in Ottawa to discuss its concerns and solution. A delegation composed of Donald Chant, Peter Middleton, Brian Kelly, and Geoff Mains of Pollution Probe at the University of Toronto, with Phil Reilly of Pollution Probe's Carleton University affiliate, met with Sharp, Greene, and Fisheries Minister Jack Davis. Pollution Probe's nine-page statement, "The Need for a Comprehensive Canadian Energy and Resource Policy," was discussed. The cabinet ministers' reactions ranged from Davis's apparent interest and Greene's indifference to what Kelly characterizes as Sharp's "very traditional, close-minded and petty" attitude. Kelly summarized the meeting by noting that "we had the opportunity to present our views and to discuss them but did not feel that we had received any commitments, or made any headway other than merely exposing them to our ideas."[22]

Despite Pollution Probe's campaign, the November talks resulted in an increase in the US oil import quota, and the two parties agreed to negotiate a free-trade policy for oil in spring 1971. Nonetheless, Pollution Probe found reason for optimism in a 19 November news report that Jack Davis had told a New York seminar on bilateral relations that the Canadian government was considering using its energy resources to encourage the Americans to work harder at cleaning up the Great Lakes, particularly with respect to the phosphate issue, which was being resolved north of the border.[23] Viewing this as proof that the federal government would come onside if it was sufficiently educated, the Energy and Resources Project urged the public to continue voicing their "dissatisfaction with the piecemeal approach which is leading us towards a Continental Energy Pact."[24]

The Energy and Resources Project switched gears in 1971. Previously it attempted to develop a groundswell of support for limiting energy exports; now it aimed to develop resource policy expertise utilizing Canada's academic and private spheres. In January, it established an energy panel composed of the ubiquitous Dr. Chant, the University of British Columbia's renowned ecological economist, C.S. Holling, and businessman Mel Hurtig. Hurtig, a founding member of the Committee for an Independent Canada, was one of the country's leading economic and cultural nationalists and a vocal opponent of continental oil integration. His inclusion in a panel dominated by academics demonstrates Pollution Probe's willingness to ally itself with other forces in the pursuit of common ends. This panel met twice in the ensuing year. A water panel was also established but soon ground to a halt as the cost of flying the far-flung members to Toronto quickly became more than the perpetually cash-strapped Pollution Probe could afford.[25]

The Recycling Imperative

Pollution Probe first mentioned recycling in October 1970 as part of the work of the Energy and Resources Project. It was portrayed as one way to help slow the exhaustion of the world's finite natural resources. Although the inherent benefits of recycling had been demonstrated during the First and Second World Wars, when salvage campaigns became a critical part of the war effort, the postwar years witnessed a mass abandonment of the practice. While there were limited contemporary examples of recycling's potential – for example, Madison, Wisconsin, initiated the United States' first municipal curbside newspaper recycling program in 1968, and recycling depots were just beginning to gain prominence in American cities[26] – it was clear that in order to address skeptics' concerns, Pollution Probe would need to undertake local demonstration projects to prove its feasibility. In early 1971, the group began focusing its energy on Toronto's telephone directories. This was a logical choice: weighing in at over 4.5 pounds each, over one million phone books were put into circulation in the city annually, and although waste paper had a relatively strong resale value, the outdated directories were regularly sent to city-operated incinerators or landfills. The project began in February, a month before the release of the updated directories. Gregory Bryce, an active member of the University of Toronto Outing Club who had been drawn to Pollution Probe after attending a series of lectures that it had co-sponsored, was hired to oversee the project, and by the end of March, plans for a Metro-wide telephone directory recycling drive were cemented. Under the organizational oversight of Pollution Probe, children from ninety area schools collected old telephone directories, which were then loaded into a forty-foot transport truck supplied by Smith Transport, as well as a number of smaller trucks provided by Bell Canada. The books were dropped off at the recently closed Don Incinerator on Gerrard Street, were baled by Metro Works Department employees, and were ultimately sold for $17 a ton to the Continental Can Company. Over the course of four school days in April, between 48,000 and 65,000 telephone directories, weighing nearly 130 tons, were diverted from the city's waste system. Bryce's project corresponded with and coordinated 171 schools, companies, and other interested parties, garnered considerable media attention, and provided thousands of schoolchildren with a hands-on opportunity to make a positive contribution to the environment.[27] In a 19 April 1971 note to the recycling drive's supporters, Bryce wrote that the "campaign has been eminently successful in developing public awareness of recycling."[28]

It was increasingly evident that recycling would figure prominently in Pollution Probe's activities, but it also happened to be an area in which the organization lacked expertise. During the summer of 1971, funding from the federal government's Opportunities for Youth program was used to hire Bryce, Clive Attwater, Sean Casey, and C. Dana Thomas to study the existing waste problem in Toronto and explore recycling technologies employed throughout North America and Europe. The project also contained a political element, with efforts being made to determine the stance of various government departments vis-à-vis recycling. Having discovered that governments were, at best, hesitant to throw their support behind recycling programs, Pollution Probe attempted to influence their policies through direct correspondence as well as newspaper, television, and radio appearances. The project also embraced public education, distributing to the general public fifty thousand copies of a recycling booklet produced

FIGURE 12   In April 1971, Pollution Probe initiated a telephone directory recycling drive in Metro Toronto. This event involved coordinating 171 schools, companies, and other interested parties, and was responsible for diverting approximately 130 tons from landfill. Pictured are unidentified Pollution Probe volunteers, surrounded by some of the collected directories. *Source:* Toronto *Telegram* fonds, ASC17110, 1974-002/161, Clara Thomas Archives and Special Collections, York University.

by the summer employees.[29] The project's 118-page final report was made available to interested parties. Detailing the project's findings and activities, the report concluded with eight pages of recommendations for federal, provincial, and municipal governments, industry, and citizen groups. These recommendations would serve as the basis of Pollution Probe's recycling policy in the ensuing years, and were designed to serve the following goals:

- change in the predominant attitude towards garbage, particularly among municipal officials and the general public

- establishment of policy by all levels of government for the ultimate recycling of all materials now considered to be waste
- recycling should constitute one element in a national energy and resources policy which recognizes the limited availability of *all* resources for both domestic use and export
- the inclusion of social costs in product prices.[30]

Upon conclusion of the project, Attwater, Casey, and Thomas returned to their studies at the University of Toronto. Bryce, the only non-student, remained on staff and became the new recycling coordinator.

Proponents of recycling found themselves in a catch-22: the City of Toronto had no interest in developing programs that might be unsuccessful, so the viability of recycling programs remained unproven. To break the impasse, Bryce led an effort to institute a weekly multi-paper collection in his home neighbourhood of Moore Park. The first project of its kind in Toronto, it was well suited for Moore Park. The tony neighbourhood boasted a strong sense of community, as evidenced by its active ratepayers association. Newspaper readership was high, providing a steady stream of material. And although Moore Park residents had long had their garbage picked up from the side or back of their homes, the proposed recycling program would require them to bundle their papers separately and leave them at curbside. As such, Moore Park would demonstrate whether the public was willing to endure slight inconveniences in order to support recycling initiatives.

The project received the support of Streets Commissioner Harold Atyeo, with the key stipulations that Pollution Probe secure a market while also demonstrating that the local residents supported the plan. Pollution Probe lined up a paper dealer who would purchase the city's collected materials and resell them to a paper mill. It also secured a guaranteed market for the dealer with the Continental Can Company. Next, the plan had to be sold to local residents. After it received a hearty endorsement at the annual meeting of the Moore Park Ratepayers Association on 17 May 1971, Pollution Probe organized a thorough information blitz of the area. Led by Bryce, volunteers went door-to-door distributing pamphlets and answering questions. Homes that did not respond to the canvassers received a follow-up visit the next day.[31] Headlined *"Residents of Moore Park, we need your help!"* the pamphlets featured basic information about Toronto's garbage problem, the need to generate less garbage and to recycle what continued to be produced, and details of the proposed plan. The pamphlet also demonstrated the city's support, including endorsements from local aldermen Paul Pickett and William Kilbourn, and a statement from Atyeo that "this plan is feasible and necessary, and [I] offer my department's support in its implementation." On the final page of the pamphlet was a questionnaire that examined the residents' interest in the project.[32] As Pollution Probe explained, it saw the Moore Park project as a necessary step in the push for municipally operated recycling pickups in Toronto:

> We are trying to develop a newspaper recycling system that will continue on a long-term basis. We do not want to depend on the temporary enthusiasm of volunteers, nor on free labour. If a small scheme works, a larger system can be developed. Ultimately, of course, we would like to see all of the city's garbage recycled.
>
> People in industry and government anticipate many obstacles to successful recycling systems. We have tried to tackle those obstacles. We will not solve the problem by talking about them, but only by confronting them in action.[33]

Statements in the media from the City Streets Department verified that it would use the results of the Moore Park experiment to determine whether a citywide collection was merited.[34] On 24 July, Pollution Probe's army of volunteers made their third visit to Moore Park, this time to collect the completed questionnaires. Almost 83 percent of respondents indicated that they were "willing to co-operate fully" with the project, while another 9 percent offered qualified support.[35]

Weekly curbside pickup began on 15 September. Just three weeks into the project, Harold Atyeo presented a report to the City Works Committee, recommending that collection be turned over to a private contractor. While collections yielded eighteen tons in the first three weeks – a figure exceeding the initial estimate of five tons per week – Atyeo pointed out that they were losing $8 per ton. Because of these losses, the program was handed over to a private contractor after the fourth week of collections. In the months that followed, the pickup continued to generate respectable results. Between 13 October 1971 and 7 June 1972, the program averaged 3.9 tons of recyclable paper per collection. In June 1972, the City of Toronto, inspired by the Moore Park program, began a municipality-wide paper pickup on a monthly basis. The monthly program, which ran from June 1972 to April 1973, averaged 175 tons per month. The city considered this underwhelming, given that it accounted for only 10 percent of the city's newsprint, but the newly founded Toronto Recycling Action Committee, a subcommittee of the Department of Public Works, came to the program's defence. Bryce, Pollution Probe's representative on the subcommittee, had long maintained that the program was needlessly complicated, to the detriment of participation levels. For example, the monthly collection was held on a different date each month. Also, since the garbage and recycling collections were handled by different trucks, paper was often mistakenly sent to landfill sites. Bryce therefore advocated that special racks be attached to Toronto's garbage trucks to enable the collection of recyclable paper during the regular weekly pickup. As a result of such critiques, additional markets were secured for the newspapers collected, and on 8 May 1973 the Committee on Public Works approved a weekly citywide newspaper pickup, to take place every Wednesday.[36]

Recycling, as a matter of waste control, was a municipal concern, but in autumn 1971 Pollution Probe turned its attention to the federal government. Arguing that the federal government should provide the impetus for recycling nationwide, Pollution Probe decided to bring a truckload of recyclables to Parliament Hill for presentation to the Honourable Jack Davis, minister of the environment. This idea, attributed to Tony Barrett, quickly evolved into the Resources Recycling Caravan, an event designed to achieve maximum media attention. With a forty-five-foot tractor trailer at its disposal, and a driver paid for by Carling Brewery, Pollution Probe scheduled an eight-day trek through Ontario, beginning in Windsor on 7 October and wrapping up in Ottawa on the 14th. In conjunction with local environmental groups, many of them Pollution Probe affiliates, the truck would stop at a series of recycling depots and pick up sorted recyclables at each location, to be deeded to Davis to fund research on recycling.[37]

As the caravan worked its way through the province, it was frequently visited by municipal and provincial politicians eager for a quick photo opportunity. Even bureaucrats joined in the limelight, as seen when John Thatcher, the provincial deputy minister of the environment, helped load the truck with recyclables. During the stop at Queen's Park, Thatcher was quick to pin responsibility for recycling on the municipalities, but he also indicated that his department was considering funding those municipalities that adopted the practice. Most of the caravan's stops were in the more populous communities of southern Ontario, such as Waterloo, where it picked up a ton of used computer cards from a local university, and Windsor, where it picked up one ton of glass and a quarter-ton of compost, but it also made stops in smaller communities such as Grimsby, where it collected a ton of tin cans, and Prescott, where the local Kiwanis Club donated 1,500 pounds of mixed paper. At each stop, Pollution Probe made the local news, promoting the idea that recycling waste was a worthwhile endeavour.[38]

The caravan's journey ended in Ottawa on 14 October. Departing Carleton University at noon, the tractor trailer and five trucks loaded with materials collected in Ottawa received a police escort to Parliament Hill.[39] With Davis scheduled to sign the deed at 1:10 p.m., Monte Hummel climbed atop a podium made of crushed soft drink cans and addressed the gathering. Noting that "we've come to Ottawa today bearing gifts," he outlined Pollution Probe's grander purpose:

> [We] have not started recycling depots or brough[t] this van to Ottawa in an attempt to take over the wast[e] disposal system but as a gesture to our elected officials who should be managing this job properly, that we want it done in a new way ... What you can see in this van represents an abundance of accessible, cheap, already processed secondary material. Where else can you find resources so close to manufacturing centres in such conveniently large concentrations? How much environmental deterioration might be avoided by recycling processed materials instead of extracting new ones? And why do this for export without insisting that foreign consumers of Canadian resources also learn to practice the principals [sic] of recycling? How many new jobs might be created by an industry as labour intensive as recycling? And how much revenue might we recover through re-use instead of just spending it on disposal[?][40]

Davis dashed Pollution Probe's hopes, however. Despite praising the environmentalists' recycling drive for achieving "something our industries ... haven't been able to accomplish," he refused to sign the deed, explaining that there was no way he could guarantee that the money would be utilized in the manner requested.[41] His refusal to sign the deed was a matter of considerable frustration for Pollution Probe. Its November newsletter explained:

> For several weeks beforehand Mr. Davis' department was made aware of our intentions of coming to Ottawa with gifts and policy recommendations. In fact the most unique aspect of the project was that we were giving a grant to the government, certainly an unusual turnabout of normal circumstances, and a tangible gesture of support for the Federal Government to get moving on recycling. The signing of the document ... was the particular gesture around which the granting ceremony was to centre.

A week before our appearance in Ottawa, as [the] Caravan was crossing the province, Mr. Davis' office called to ask for a change in the document and wording which we agreed to. The ceremony was scheduled for 1:10 p.m. At 11:00 a.m. the minister's office called to say Mr. Davis would not sign. CRISIS! What to do. Well, we w[e]nt ahead with everything as planned except at the scheduled time in the ceremonies for the signatures, we signed and Davis did not.[42]

Pollution Probe openly speculated that Davis's about-face could be attributed to a fear of raising expectations of action that the government was not prepared to meet, pressure from primary resource industries, whose extraction business would be negatively affected, and fear on the minister's part that the signed deed might be misconstrued as a legally binding agreement to support a national recycling initiative. Adding insult to injury, the final act of the Resources Recycling Caravan failed to garner the anticipated media attention. Despite orchestrating a memorable publicity stunt, it had the misfortune of occurring on the same day that a federal budget announced major cuts to personal and corporate income tax rates.[43]

Growing the Movement at Home and Nationwide

Pollution Probe's encouragement of affiliate groups showed its understanding of the need to grow the environmental community. In the latter half of 1970, it fostered the development of four environmental organizations. The first was CAHE. Initially set up by members of Canada's conservation community, the organization fell into dormancy in the 1960s. It was relaunched with the blessing of its founders at a national convention in September 1970, and, headed by Pollution Probe's Peter Middleton, it now served as an umbrella group representing ENGOs from nine of the ten provinces. These organizations were a disparate lot, ranging from relatively large groups with paid staff to small, volunteer-driven groups scattered across the country, and separatist-led groups in Quebec. The sole purpose of CAHE was to create the infrastructure necessary to get the maximum funding available through federal student employment programs such as Opportunities for Youth and the Local Initiatives Program.[44] As Middleton notes wryly: "That was the glue that made for national unity, as it often has been in the history of the country."[45] Although it received little attention — credit for projects went to the local groups rather than the national body — CAHE was nonetheless quietly effective. In the summer of 1971, for example, it received funding for projects that employed almost seven hundred Canadian students, including forty-two in Toronto who worked for Pollution Probe. CAHE lasted for three years, dissolving when funding dried up.[46]

September 1970 also gave rise to the Council Organized to Protect the Environment (COPE). COPE was designed to mobilize existing community, social, service, religious, and financial groups in a role complementary to that of Pollution Probe. Organized with the help of Rob Mills, its stated membership included the Anglican Church, the United Church, the Central Branch of the Unitarian Church, the Toronto East Presbyterian Church and Toronto West Presbyterian Church, the Council of Catholic Men, the Holy Blossom Temple, the Toronto Council of Jewish Brotherhood, the Central and Scarborough Home and School Councils, the Toronto B'nai Brith, the Metro YWCA, the National Council of Jewish Women, and the Junior League. COPE's first project was a citywide door-to-door survey with seven questions concerning the respondents' lifestyle choices and their environmental implications, and it utilized volunteers from Metro high schools. Viewing COPE as a valuable link to the broader community, Pollution Probe established a liaison with it and invited it to send a representative to Pollution Probe's weekly coordinators' meetings. COPE tapped into the growing desire of established groups in Toronto to support Pollution Probe, but proved to be superfluous as members of the public could more easily direct their support to the better-known ENGO. With nothing new to offer the environmental community, COPE was doomed to a short lifespan, with no evidence of its existence beyond June 1971.[47]

Much more successful was Pollution Probe's foray into law. It became apparent that the legal system was a great untapped resource; consequently, Barrett and Middleton began recruiting interested parties from Toronto's law schools. As Alan Levy, one of the law student recruits, explains:

> The concept was to create a public interest law clinic that could provide support for environmental groups like Pollution Probe that needed expertise (there was very little at that time in the private bar) at little or no cost ... At the time, [Pollution] Probe was receiving numerous calls from people living in Ontario and beyond with environmental concerns and problems, and wanted a legal team mobilized to be able to assist them.[48]

This led to the birth of the Environmental Law Association, renamed the Canadian Environmental Law Association (CELA) in 1972, as well as the Canadian Environmental Law Research Foundation (CELRF), a sister organization established to conduct legal and policy research.[49]

The practice of environmental law posed particular challenges. First, the concept was still in its infancy. The Environmental Defense Fund, an American group formed in 1967, used scientifically backed litigation to push for a ban on DDT.[50] Barry Stuart was offering Canada's first environmental law course at the Osgoode Hall Law School beginning in September 1970, but there were no professional associations or firms associated with the practice. In fact, as reflected by law professor D. Paul Emond, one of Stuart's students in the initial offering of the course, there was no such thing as environmental law. Rather, "there was optimism that, with enough imagination, a good lawyer (or law student) could cobble together tort, property, and perhaps criminal law to stop, or at least severely curtail, any pollution problems. If that was not enough, then the hope was that strong advocacy would persuade governments to pass effective environmental protection legislation."[51]

In the early days, CELA's work was primarily handled by articling students from the University of Toronto's Faculty of Law. With no funds at their disposal, the volunteers would meet at their homes after class and divide up complaint reports collected by Pollution Probe's Pollution Complaint Service.[52] In the summer of 1971, CELA received a federal grant that provided five full-time employees with a base salary of $70 per week to pursue their work. Office space came from a familiar source: Dr. Donald Chant and the Department of Zoology at the University of Toronto. After a number of short-term hires funded by an assortment of government grants, CELA hired its first full-time general counsel, David Estrin, in December 1971. Formerly employed at a general practice law firm, Estrin saw his annual salary of $10,000 halved after his move to CELA. "Fortunately," he recalls, "my wife at that time ... had a regular job so we were able to get by."[53] Given that the organization lacked stable funding and continued to survive from grant to grant, there was no assurance that Estrin's salary would be met. Financial difficulties continued to hound CELA, and in 1975 several members of its board of directors signed personal guarantees in order to keep the organization afloat. The prognosis improved greatly the following year when it began receiving support from Legal Aid Ontario amounting to $2,000 a month.[54]

CELA's first high-profile court case arose out of an imbroglio over excavations at the Sandbanks Provincial Park in Prince Edward County. As the name implies, the park, established in 1957, was noted for its sand dunes, some of which stood over a hundred feet above the neighbouring shore. The Ontario Department of Forestry began a tree-planting project in an effort to contain the sand dunes, which shift naturally as much as forty feet a year. Two years later, it was discovered that thousands of trees had been planted on sixteen acres of neighbouring land belonging to the Lake Ontario Cement Company. Despite the company's protests that the trees made it uneconomical to continue excavating sand from its property, the provincial government hesitated to remove the trees for fear of leaving nearby farmland unprotected from the shifting sand. In 1967, following a long-standing court case that solved little, Attorney General Kelso Roberts granted Lake Ontario Cement a Crown lease to sixteen acres of the provincial park. The deal, which cost a token dollar per year, was good for unlimited excavation for seventy-five years.[55]

In 1971, public concern began growing over the extent of dunes being destroyed by Lake Ontario Cement. Such sentiments were dismissed by area MPP Norris Whitney, who scoffed at the "increasing numbers of urban people who have scant consideration for the interests of local citizens in those areas where they take their brief vacations."[56] In a series of letters to the *Globe and Mail*, Whitney noted that Lake Ontario Cement provided well-paying jobs, something that was in short supply in the region. Harold Cantelon, a local parks supervisor for the Department of Lands and Forests, argued that the excavation deal would benefit tourism, as it would create sixteen additional acres of white sand beach at no expense to the public.[57] As pressure mounted, rumours began swirling that the provincial government was negotiating to buy the land back, while local Tory James Taylor, a candidate in the forthcoming provincial election, openly discussed alternative sites for the quarry. Despite a flurry of discussion prior to the election, the issue died down in the ensuing months, prompting an editorial in the *Globe and Mail* to ask: "Didn't Government officials claim two months ago that they were working feverishly to find the company a new site so the dunes could be saved?"[58]

The lack of progress was an irritant to the anti-extraction forces, as the sand dunes were being removed at the rate of eighty thousand tons annually, meaning the entire sixteen-acre parcel would be flattened within fifteen years.[59] CELA developed a plan to sue the Ontario government. With Pollution Probe's Larry Green acting as plaintiff, on 4 May 1972 CELA served Attorney General Dalton Bales with a formal notice of claim stating that the province had breached the public trust by allowing Lake Ontario Cement to excavate a site protected under the Provincial Parks Act, which,

CELA argued, necessitated that the lands be maintained "for the benefit of future generations." The suit also argued that Lake Ontario Cement's failure to obtain a licence under the Beach Protection Act also rendered the company's actions illegal. CELA's notice, filed by David Estrin, gave the province sixty days to halt the excavations.[60] When the government still did nothing, a formal suit was brought forward on 8 August 1972 by CELA on behalf of Green, charging the government with a breach of trust for failing to maintain the Sandbanks for "healthful enjoyment and education," as required under the Provincial Parks Act.[61] On 5 July, an estimated 150 locals and vacationers staged a ninety-minute picket, preventing trucks loaded with sand from departing the provincial park. The protest was sparked by complaints that Lake Ontario Cement's noisy excavation process was awakening those in the tourist destination at 5:30 a.m., which owners of the nearby Sandbanks Beach Resort claimed violated an earlier agreement that the company would halt excavation during the months of July and August.[62] CELA also filed seven charges under Section 14 of the Environmental Protection Act on behalf of Agda Rayner, a Toronto secretary who had been staying at the Sandbanks Beach Resort. Described in the *Globe and Mail* as "an unprecedented application of the mischief section of the Criminal Code," CELA's application consisted of charges against Lake Ontario Cement and the onsite heavy equipment operator, Triad Truckways, for "mischief involving willful interference with the enjoyment of private property," to accompany charges against each for impairing the environment at the provincial park and at the resort on 4 July and again on 7 July.[63]

July also saw the release of a Department of Natural Resources report on the Sandbanks prepared by Dr. Walter M. Tovell, a geologist and associate director of the Royal Ontario Museum. Tovell rejected environmentalists' concerns that the excavations were causing irreparable damage to the provincial park, noting that the dunes in question covered just 1 percent of the 1,802-acre park. Furthermore, he argued that even after a complete excavation of the site, sand dunes would return within fifteen years, due to their fluidity. Tovell did acknowledge the politically sensitive nature of the issue, and recommended that Lake Ontario Cement should expedite the excavations in order to reduce tensions. Two months later, CELRF and Pollution Probe released a collaborative report refuting Tovell's findings. They argued that his claim that the sand dunes would naturally regenerate was baseless. Noting that the free flow of sand had been halted by the government's reforestation plan, Pollution Probe's onsite research revealed a series of large depressions approximately eighteen feet deep, which had flooded and were filling with algae, swamp grass, and marsh weeds. These depressions, it was argued, altered the site's fundamental landscape. It was further alleged that Lake Ontario Cement had purchased the property in question on 21 October 1958, after it had been reforested, and that the provincial government had agreed to the land swap only in order to avoid a protracted lawsuit. This report also drew attention to potential links between the provincial government and Lake Ontario Cement, whose board of directors included former cabinet minister Michael Starr, while the company with a controlling interest in Lake Ontario Cement, Denison Mines, had a board of directors that featured a number of prominent Conservative supporters.[64]

Both sets of charges were heard in October. Rayner's criminal charges were heard in Picton, with Aubrey E. Golden, Estrin's former employer, handling the prosecution on behalf of CELA. Before any evidence could be entered, however, the case was thrown out of court on the grounds that the provincial Environmental Protection Act was invalid: air and noise pollution were determined to be matters of federal jurisdiction. The civil case, brought before the Ontario Supreme Court, was similarly thrown out on the grounds that "breach of public trust" was not acknowledged as suitable grounds for a trial.[65]

While those involved in CELA's action were understandably disappointed – particularly Green, who was found liable for the court costs of both Lake Ontario Cement and the government – it was not a complete loss. By raising the issue, CELA was able to focus public attention on the problem, and even though it lost the case, it ultimately forced the government's hand. In the aftermath of the charges in the Rayner case, Lake Ontario Cement halted excavations at the Sandbanks Provincial Park, and on 21 March 1973 the provincial government announced that it was cancelling the lease. Estrin credits the Sandbanks case with inspiring a change in the way the Ontario Ministry of the Environment operated. The ministry had been hesitant to enforce its regulations, for fear that it would end up on the losing end of a court trial. In the wake of this case, it lost much of its reluctance and eventually developed an investigation and enforcement unit.[66] In a strange twist of events, the Sandbanks criminal case, which saw the judge throw out the province's jurisdiction over environmental matters, resulted in a collaboration between CELA and the provincial government. Estrin was working at his makeshift office in the Ramsay Wright Zoological Building when he received a telephone call from Environment Minister James Auld inviting him to lunch. Upon arriving at the upscale Bay Street restaurant, Estrin discovered that Auld wanted to discuss the Sandbanks case. He recalls: "It began to dawn on me why they're so concerned. If the judge's ruling was not reversed, they would be without a job because there couldn't be a provincial minister of the environment, [and] there wouldn't be any jurisdiction. [laughs]" Estrin advised Auld to file an appeal via the Attorney General's office, and to put Morris Manning on the case.[67] The ministry agreed, and on 16 March 1973, Justice John O'Driscoll of the Ontario Supreme Court upheld the province's jurisdiction over environmental protection, explaining that "pollution is, or should be, the concern of each person in Ontario and, indeed, throughout the world."[68] As this demonstrates, despite losing the Sandbanks case, CELA had accomplished its initial aims and established itself as a credible organization, particularly in the eyes of the ministry.

Pollution Probe's success in the early 1970s did not go unnoticed by Canada's other ENGOs. Its high media profile and fundraising prowess resulted in a steady stream of requests for advice. In response, Pollution Probe dispatched its staff to hold workshops with ENGOs across the country, including the Halifax-based Ecology Action Centre and British Columbia's Scientific Pollution and Environmental Control Society. While these workshops tended to emphasize Pollution Probe's approach to fundraising, its organizational structure and relationship with the media were also commonly discussed. The meetings appear to have been particularly meaningful for the Ecology Action Centre, which began to approach government, private corporations, and foundations for support while adopting a project structure similar to that utilized by Pollution Probe.[69]

Maturation of Pollution Probe's Structure

Pollution Probe's decision-making apparatus failed to keep pace with its expansion in terms of staff and the scope of its projects. Day-to-day operations were handled by the paid staff, while important issues were discussed at general meetings, where each member had an equal say and great pains were taken to reach a consensus before acting on an issue. Pollution Probe managed to function without a designated leader in the early days, but the increasing scale of the operation, in terms of staff and budget as well as range of activities, led to calls for the creation of an executive director position. Support for this position was not universal as many members, primarily volunteers, wanted to maintain the status quo. These members looked to the model of cooperatives, which were popular within the New Left throughout the 1960s and 1970s. As Joyce Rothschild and J. Allen

Whitt explain in *The Cooperative Workplace: Potentials and Dilemmas of Organisational Democracy and Participation*, the cooperative model could be effective when utilized by a small group sharing a common approach and ends.[70] Pollution Probe, however, was already showing signs of segmentation. While all members ultimately desired a healthier environment, the immediate priorities of staff working on recycling differed from those of staff working on energy and resource issues and those who were more interested in public education efforts. Without a designated leader, meetings often turned into marathon debates over the allocation of resources as well as the general direction of the ENGO. These debates typically ended inconclusively and were extremely frustrating for many members.[71] As Monte Hummel explains: "I can remember one meeting where a member put his fist through the wall saying 'This is hopeless. We're not going anywhere. We've got to make a goddamn decision here ... It [the leaderless format] became counterproductive and dysfunctional."[72] The resulting efforts to impose a hierarchy came to a head on 25 May 1971 when Paul Tomlinson, one of the initial four employees hired in 1969, announced his resignation. Noting that he could not "sit idly by and watch the demise of an organization which I have helped to build," he argued the need to hire an executive director in order to prevent the group from becoming "bogged down by its internal dynamics." Tomlinson further stressed that the ENGO "can no longer afford to function as an uncoordinated family compact. The 'do-it' philosophy still holds true, but the lack of a traditional hierarchy which has worked with a group of three or four, will not work now with fifteen and certainly will not work with 50."[73]

The need for an executive director was also argued by Peter Middleton who, as the coordinator in charge of internal Pollution Probe communications and the liaison with CELA and ENGOs across the country, was already shouldering much of the burden associated with such a role. On 28 May 1971, he issued an ultimatum:

> The time has come to resolve once and for all my status at Pollution Probe and especially in light of Paul's resignation and his reasons (which I for the most part agree with). The time has come for me to either exercise fully the responsibilities which people would sometimes willingly pass on to me or to remove myself completely from the scene ... It is unreasonable to expect every co-ordinator to spend time and effort trying to keep abreast of what everyone else is doing. [Pollution] Probe needs someone who will be in touch with everyone and ... will be able to bring the co-ordinators together

when their activities suggest that this is necessary and plug the skills of various people into various projects at different times.[74]

Middleton argued that providing the organization with a coherent direction and channelling its members' energies would be a natural extension of his existing role with Pollution Probe. Furthermore, he shared Tomlinson's fears that the ENGO was on the verge of becoming dysfunctional. The idea of having one staff member elevated in status was controversial at the time; however, the argument in favour of its utility eventually won out, and in the summer of 1971, Peter Middleton assumed the position of executive director.[75]

Pollution Probe's adoption of a hierarchical model coincided with its breakthrough at Queen's Park. Although the ENGO understood the necessity of accessing the corridors of power and submitted a number of briefs to Premier John Robarts, there are no records of any personal meetings with him. The closest it managed to get appears to be a meeting held in June 1970 when a group of environmentally conscious schoolchildren, accompanied by members of Pollution Probe, met with the premier to discuss non-returnable bottles.[76] Pollution Probe had more success when Bill Davis became premier. A lawyer by profession, Davis was first elected to Queen's Park in 1959 at the age of twenty-nine as the representative for Peel. Between 1962 and 1971 he served as Ontario's minister of education. Pollution Probe was openly skeptical of Davis's environmental pedigree, given his "disgraceful fourth" place finish in its February 1971 environmental survey of the Progressive Conservative leadership candidates,[77] but the two parties eventually forged a strong relationship. Despite eschewing the label "environmentalist" in a 2008 interview, Davis added that "I considered myself one concerned about the environment, and one who listened to others and endeavoured to do something about it." This concern, piqued by water issues in the Georgian Bay, where he had his summer home, opened the door to consultation with Pollution Probe, which he considered credible due to its academic connections and especially its relationship with Donald Chant. As he explains: "We developed a relationship with them that I think most of them would say was ... very cordial. We didn't agree with everything ... but I think the relationship was one that was fairly productive."[78]

Pollution Probe's relationship with Davis was no doubt aided by the fact that his chief policy adviser, Eddie Goodman, was on the CELRF Board of Directors. "Being very canny politicians," Middleton explains, "they could see our appeal and they could see we could deal with each other for mutual benefit."⁷⁹ This connection was in place by 3 August 1971, as the second page of the Toronto *Star* carried a story about a half-hour meeting between Pollution Probe representatives and Davis in the latter's office, after which the young environmentalists "invited him out to the Queen's Park north lawn for a lunch of sandwiches and milk." The article noted that Pollution Probe's chief concerns were the difficulty its members were having accessing government information pertaining to the environment, and the apparent hesitation of government officials to speak with the organization. "There has been a certain lack of confidence, to put it mildly," Middleton informed the media. As the article intimated, Davis agreed to Pollution Probe's request for increased access to government information.⁸⁰

The Amchitka Interlude

In the fall of 1971, the United States Atomic Energy Commission planned to detonate a nuclear device underneath the remote island of Amchitka. Part of the Aleutian Islands chain, bounded to the north by the Bering Sea and to the south by the Pacific Ocean, Amchitka had played host to previous nuclear tests in 1965 and 1969. Organized opposition in Vancouver, British Columbia, stoked by anti-American sentiment and fears of the potential environmental fallout, resulted in the Don't Make a Wave Committee – the future core of Greenpeace – attempting to navigate an eighty-foot halibut seiner named the *Phyllis Cormack* into the detonation zone in an effort to halt the test. The vessel was intercepted by the US Coast Guard and turned around, but not before the activists on board captured the hearts and minds of citizens around the world with their direct action tactics.⁸¹

Opposition to the blast was particularly strong in Canada. Parliament passed a motion against the detonation, while a petition calling for its cancellation, drafted by authors Pierre Berton and Charles Templeton, garnered 177,000 signatures. Thousands took to the streets to show their disapproval, with 8,000 encircling the American consulate in Toronto and 5,000 protesters from the two countries halting traffic on the Ambassador Bridge, which connects the cities of Windsor, Ontario, and Detroit, Michigan.⁸²

It was in this politically charged climate that Pollution Probe engaged in uncharacteristically radical behaviour. Ann Love, the information coordinator at the time, recalls that the organization "decided to do something short and sweet" to voice their opposition to Amchitka. In the five days leading up to the 6 November detonation, staff at Pollution Probe sent a daily "gift" of a dead item to the American consulate in Toronto. Each item was sent by taxi and was accompanied by a signed letter from the ENGO. Things began rather innocently on Monday, 1 November, with a bouquet of dead goldenrods. The following day a wren that had died when it flew into Love's window was delivered on a silver platter. On Wednesday, it was a dead rabbit, lying atop a bed of leaves, and Thursday saw the delivery of a fetal pig in a glass jar. On the fifth day, Pollution Probe sent a frozen pig's head along with a note proclaiming that "the pig is a legendary symbol of greed. He devours indiscriminately. But what he doesn't realize is that his uncontrollable appetite also hastens his own death. Stop the blast!" Two days after the detonation, on Monday, 8 November, Pollution Probe sent its final gift, a bottle of dead fish with a note that read "You may have killed only fish. Don't take another chance. Don't make another Amchitka."[83]

Eight hours after sending the final gift, Pollution Probe was visited by the taxi driver. Furious at having been detained and interrogated at the consulate over suspicions that his delivery contained a bomb, he demanded $20 as compensation for his ordeal. Love sympathized and ensured that he was paid.[84] Dr. Chant, who had earlier defended the deliveries, noting that "it may be offensive to some people's taste but the Amchitka test is an obscenity which should offend everybody's taste," was visited by members of the Royal Canadian Mounted Police investigating what they thought was a possible terrorist action.[85]

As it turns out, the national furor over the Amchitka nuclear detonation had caused Pollution Probe to stray from its standard mode of operations. Love realized that she had stepped outside her personal comfort zone only after being confronted by the angry taxi driver; in addition, her name, as one of the leading figures in the campaign, featured prominently in the ensuing media coverage, infuriating her father. Despite the protests of thousands, the detonation took place as scheduled on 6 November. In a final twist, Love received a parcel at the Pollution Probe headquarters two weeks later: "I was sitting at a meeting and it was a reasonably large envelope and it was sort of bumpy." Curious, she opened the envelope, which contained the silver platter that had been sent, along with a dead wren, to the American consulate on 2 November. Eventually, the passage of time caused her father, famed physician and stroke researcher Henry Barnett, to embrace his daughter's participation in the campaign, and he now boasts of it as one of her great accomplishments.[86]

Conclusion

Pollution Probe evolved significantly during the period examined in this chapter. It grew from an ENGO with a skeletal staff to one with sixteen employees in September 1970. The addition of an executive director provided a sense of direction and oversight that it had previously lacked. Its objectives expanded beyond combating air and water pollution to include energy and resource issues, as well as the creation of additional environmental organizations. Although by no means averse to generating publicity through protests, Pollution Probe showed an increased proclivity to address its concerns through existing bureaucratic channels.

Insight into Pollution Probe's evolution can be gleaned from the work of Paul Pross. A scholar of Canadian pressure groups, Pross has argued that they exist in four increasingly sophisticated forms. The most simplistic organizations, which he terms "issue-oriented" groups, are characterized by a narrow focus on one or two issues, fluid membership, limited organizational cohesion, a focus on publicity-seeking events, a "considerable difficulty in formulating and adhering to short-range objectives," and a confrontational approach towards officials. The next stage is the "fledgling" group, defined by multiple, closely related objectives, a small paid staff supported by membership, the utilization of briefs to public bodies, and a transition from a purely confrontational relationship with officials to somewhat regular contact.[87] In the period immediately following the CRTC hearings into *The Air of Death*, Pollution Probe, like GASP, was an issue-oriented group. Its transformation into a fledgling group was completed when it hired its first four staffers during the winter of 1969-70. More complex still is the "mature" group, which features broadly defined objectives, a staff that is, at least in part, professionally trained, regular contact with officials, and a transition from submitting briefs to using the media for public relations purposes, including the use of image-building advertisements – an accurate description of the state of Pollution Probe in 1971.

Pollution Probe would continue to expand in size and scope in 1972. While it would not attain the status of an "institutionalized" group – according to Pross such groups, which require extensive financing, were rare in Canada during this period – Pollution Probe's staff would experience yet another period of exponential growth. As will be seen in the following chapter, this would be the high-water mark of Pollution Probe's existence.

4
Probe's Peak

ON 30 JUNE 1972, the *Globe and Mail* published a letter to the editor by Frank Summerhayes, president of the Urban Development Institute – Ontario, voicing his disapproval of a recent Pollution Probe publication. As he stated at the outset: "It is distressing to see Pollution Probe damage its credibility by publishing a naive document on urban redevelopment titled Rules of the Game." Describing the publication as "a harangue against all profit-making organizations," he critiqued Pollution Probe for "clearly setting out to add to the present state of political polarization over growth and development" in Toronto.[1] The sixty-page *Rules of the Game: A Handbook for Tenants and Homeowners* was a marquee publication of Pollution Probe's Urban Team. At the time of its launch, it was described as being about "the deceptive tactics of developers, the carelessness of nearly all City politicians, the inaccessability [sic] of information, [and] the decision-making process that depends more on money than the wishes of Toronto's citizens."[2] Covering such basics as tax assessment and collection, the way City Hall plans its budget and decides which services to provide, and the way that areas are zoned, the handbook made some bold statements. Those living in the affluent neighbourhood of Rosedale, it was argued, received triple the quality of services compared with those living in the nearby working-class Grange Park or Riverdale, despite paying an equal tax rate. The publication claimed that rich Torontonians, defined as those earning an annual salary of $12,000 or more, lived in areas with a disproportionately high level of parkland. It also argued that commercial high-rise buildings failed to pay the taxes necessary to cover the cost of services provided by the municipality.[3]

It was to be expected that Summerhayes, a representative of the real estate development and construction industries, would critique the message of *Rules of the Game*. However, the publication also raised the hackles of Tony O'Donohue. Although renowned for his anti-pollution agenda, the municipal politician and former executive director of the moribund GASP was also a keen advocate of city development. As he explained at the time of the publication's launch, Pollution Probe's foray into development issues threatened to undermine the credibility it had earned through its other work.[4] Such feelings, it turns out, were also found within Pollution Probe itself, as not all members could see the direct relevance of this work to the organization's broader agenda of environmental advocacy.

The minor controversy surrounding *Rules of the Game* is emblematic of Pollution Probe's continued evolution in the period between 1972 and 1974, which proved to be the high-water mark in the ENGO's history. During the Christmas 1971 holidays, executive director Peter Middleton undertook a major restructuring of the organization designed to streamline operations. Having hit the threshold of twenty employees, a team model was instituted, with a coordinator responsible for each team. The team leaders, together with the executive director, formed a management committee responsible for decisions related to Pollution Probe's day-to-day operations. The "team" years showcased a diversity of outlooks and emphases within the ENGO. In addition to its established interest in air and water pollution, as well as energy and resource policy, Pollution Probe began addressing land development policy and matters concerning the built environment of the inner-city population.

This chapter examines Pollution Probe's "team" years. In one respect, this was a time of great creativity within the organization. At the same time, however, differing opinions of Pollution Probe's priorities exposed competing conceptions of what constituted an environmental issue.

Education Team

The most fundamental of Pollution Probe's new groupings was its Education Team. Headed by Monte Hummel, the team traced its roots to the ENGO's earliest days, when it sent speakers to schools and community groups in an effort to spread the message of environmental action.

FIGURE 13   From the time of its founding through the early 1970s, Pollution Probe sent speakers to schools and community groups in an effort to promote environmental action. Pictured here is Pollution Probe member Bob James (foreground), speaking to secondary school students during a June 1970 pollution conference. *Source:* Toronto *Telegram* fonds, ASC17147, 1974-001/365, Clara Thomas Archives and Special Collections, York University.

The value of this work was acknowledged in March 1971 by the Ontario Education Association when it honoured Pollution Probe with the prestigious Greer Award, bestowed annually on a group or individual that made an outstanding contribution to education in the province. Due to this tremendous service, in November 1970 the Metro Toronto School Board granted Pollution Probe $16,000 for past work. Much of the money was used to create and distribute teachers' kits containing lesson plans that incorporated environmental education, suggested reading lists, and advice on forming school-based environmental action groups. Supplied for free to each school in Metro Toronto, these kits sparked controversy as they included a birth control handbook to accompany information on global overpopulation and the resulting strain on natural resources. Although the School Board had noted at the time of the grant that it would consider providing Pollution Probe with a further $24,000 for its work in 1971, some schools threatened to block the grant application unless the Education Team removed the birth control handbooks. In an effort to maintain its independent status, the Education Team withdrew its application for the 1971 grant.

By 1973 the team had shifted its focus from merely providing speakers to pushing the Ontario education system to adopt a year-round program of environmental studies for all grade levels. Individual teachers developing their own curricula had complained of a sense of isolation from others doing similar work, as well as difficulty in keeping up-to-date on environmental issues. The Education Team launched a bimonthly newsletter, *Environmental Education*, to enable dialogue between educators, highlighting their successes and challenges, and environmentalists. By the end of the year, this newsletter had eight hundred subscriptions from across the province. *Environmental Education* continued publishing, albeit with an increasingly erratic schedule, through 1976.[5]

ENERGY AND RESOURCES TEAM

The Energy and Resources Team was a continuation of the Energy and Resources Project. Headed by Brian Kelly, it focused much of its efforts on provincial hearings concerning energy matters. On 11 February 1972, the team delivered a nineteen-page brief to Task Force Hydro, which was commissioned by the provincial government to study all aspects of Ontario Hydro, including its "functions, structure, operation, financing and objectives with the aim of making recommendations which will assure the quality and quantity of its services to the public in the future."[6] The team's brief was particularly critical of Ontario Hydro's rate structure, noting that the unit price decreased as energy consumption increased. The team viewed this as a reward for inefficient energy use, and further pointed out that Ontario Hydro had been promoting the use of energy-inefficient electric space and water heaters. It recommended that Ontario Hydro replace the existing rate structure with a marginal cost pricing system, in which power users would be charged the full cost. It also recommended that Ontario Hydro adjust its research, development, and advertising programs to "reflect the theme of energy conservation and the most efficient use of natural resources." The team's brief also addressed the fact that Ontario Hydro had failed to account for its external costs of operating, such as the cost of pollution to property values, wildlife habitats, and human health. Instead, Hydro's costs "are passed on in hidden ways to society (the 'polluted-upon') and to the environment which is treated as a waste disposal sink and a 'free good.'" These costs, the team argued, should be incorporated into Ontario Hydro's rate structure, in order to ascertain the true cost of energy. Finally, the decision-making apparatus at Ontario Hydro came in for critique. Describing Ontario Hydro as "a self-perpetuating, self-justifying, autonomous bureaucracy that is largely unaccountable to the public or the government," the brief noted that there was little, if any, public consultation in matters concerning "power plant sites, transmission line locations, plant types and rate changes." The brief recommended that a regulatory board, consisting of experts from medicine, ecology, engineering, law, sociology, and economics, as well as representatives of the locale directly affected, be established to "consider all proposed major actions of Ontario Hydro (and other energy-related industries) with authority to approve or reject the proposals" and to "review operating plants once every 5 years with authority to order modifications or complete close-down of such plants." After examining each proposal for its environmental impact, societal impact, and technical feasibility, this regulatory body would then open the proposal to public hearings.[7]

Pollution Probe's expertise in energy matters was recognized when it was given a seat on the Advisory Committee on Energy (ACE) that had been appointed by the provincial government "to undertake a comprehensive review to ascertain Ontario's future energy requirements and supplies and to recommend policies and means to ensure that these requirements are met."[8] Its representative on the committee was Dr. Henry Regier, a member of its Board of Advisors. Pollution Probe submitted its brief on 18 July 1972. The brief focused on the need to base Ontario's energy policy on the basic principles of energy conservation and efficiency, recognition "that exponential growth in resource consumption cannot continue indefinitely in a finite world," and the creation of an Energy Regulatory Board to oversee the energy industries. The proposed board would enforce the aforementioned environmental policies, and it "would require a comprehensive environmental impact statement, a broad cost-benefit analysis and the fullest possible public involvement before making a decision on any major proposal by the energy industries."[9] The brief also made specific recommendations regarding transportation, including encouraging energy-efficient modes of shipping, such as train and boat, making public transit free, and phasing in a ban on car traffic in downtown areas throughout the province. Recommendations for consumer reform included the promotion of long-lasting and easily reparable products and a ban on any advertising that attempted "to induce an artificial demand for a product." The brief urged the province to promote home heating with natural gas and to discourage the use of electric heating.[10]

Pollution Probe's work on Task Force Hydro and ACE emphasized Ontario Hydro's need to promote energy conservation, the incorporation of real cost pricing in its rate structure, and the democratization of its planning process, but the final reports failed to reflect these ideas. The most dramatic change advocated in Task Force Hydro's five reports, released between 15 August 1972 and 29 June 1973, was the re-establishment of Ontario Hydro as a Crown corporation operated by a board of directors.[11] The third report, which focused on nuclear energy, was predicated on the understanding that demand would continue to grow in the province and that it would be Ontario Hydro's responsibility to increase generating capacity.[12] Although it was suggested that Ontario Hydro should increase the transparency of its planning process and that electricity rates should reflect the cost of production, the Task Force Hydro recommendations did not go as far as Pollution Probe's.[13] Meanwhile, the ACE report contained some talk of conservation methods, but it accepted the premise that Ontario's energy consumption would continue to grow unabated, and projected that its energy requirements would grow two and a half times by 1990.[14] Furthermore, it stated that the province must prepare for the number of cars in Toronto and Hamilton to double in the same period.[15] All of this was indicative of the prevailing notion that the Canadian economy could be built on the availability of an ever-increasing supply of affordable energy. Henry Regier wrote a one-page minority report, focusing on the need to change the dominant approach from demand orientation to supply-side orientation, and stating that "Pollution Probe explicitly dissociates itself from all those parts of the report that follow the assumptions that high growth rates in energy consumption will continue for several decades."[16] Largely ignored at the time, Pollution Probe's recommendations would seem prescient come autumn 1973.

The Energy and Resources Team also engaged in matters of national concern. The late 1960s saw considerable exploration of the Canadian Arctic's energy potential. Spurred by generous tax incentives, ninety trillion cubic feet of natural gas and six billion barrels of oil were discovered in the Beaufort Sea. In order to bring all this to market, a Mackenzie Valley Pipeline was proposed, which raised concern among environmentalists who feared the ecological implications, as well as among Canadian nationalists who loathed the prospect of American conglomerates extracting the oil from Canadian territory for use in the American market.[17]

Pollution Probe played an early and vital role in organizing opposition to the Mackenzie Valley Pipeline. Early in 1972, the Energy and Resources Team held a series of meetings that brought interested parties together to share information about northern ecology and Canadian Arctic Gas, the consortium behind the project, as well as to brainstorm ways to block the pipeline's construction. "At this early stage," historian-activist Robert Page writes, "[Pollution] Probe played an essential role as a clearinghouse for ideas and analysis."[18] In March 1972, the Energy and Resources Team issued its first policy statement on the issue, "Freeze the Arctic," which challenged the advisability of northern development.[19] The accompanying report featured five key points. First, noting that little was known about the Arctic environment, the team called for a two-year moratorium on all new northern energy projects in order to allow time for the appropriate studies to be conducted. Second, it argued that any proposals concerning exploration or development in the Arctic should be vetted by the Aboriginal populations affected to ensure their continued ability to live off the land. Third, there was criticism of the administration of Canada's northern territories, which centralized a vast array of powers in the Department of Indian Affairs and Northern Development. Pollution Probe sought to remedy this by transferring responsibility for pollution control to the Department of the Environment and distributing responsibility for northern development and Native affairs into separate departments. Fourth, the report pointed out that in 1970 the federal government had passed two acts to control pollution in the Arctic but that neither was enforceable as it stood. It argued that these pieces of legislation, the Arctic Waters Pollution Prevention Act and the Northern Inland Waters Act, should be amended immediately. Finally, the report alleged that Arctic development was occurring without the direction of any clear policy, and argued that "more information should be made available to the general public and an official long-range comprehensive northern policy must be established and implemented after full public hearings."[20]

Pollution Probe's efforts to secure public hearings for the Mackenzie Valley Pipeline resulted in a joint proposal with the Canadian Arctic Resources Committee, the Canadian Wildlife Federation, and the Canadian Nature Federation. Submitted to the federal government in June 1973, the proposal emphasized the need for broad-based hearings that considered the social and ecological impact of the development. Since public interest groups lacked the financial resources necessary to develop credible cases, given the requirement for research, transcription, legal and witness fees, and accommodations, the proposal made the case for government funding of public interest groups to ensure that hearings were as balanced as possible.[21] Pollution Probe's campaign, in conjunction with the opposition raised by its anti-pipeline allies, played an important role in convincing the federal government to commission the Mackenzie Valley Pipeline Inquiry, headed by Justice Thomas Berger, in March 1974. Although Pollution Probe continued to express concern over the proposed development, a lack of expertise in northern matters, coupled with a lack of available funding, led to the Ottawa-based Canadian Arctic Resources Committee's assumption of the role of chief critic before the inquiry.

The fiscal year ending 30 September 1973 saw roughly one-quarter of Pollution Probe's budget dedicated to the Energy and Resources Team.[22] World events shortly thereafter ensured that its role would only increase. On 6 October 1973, Egyptian and Syrian forces invaded Israeli-held land in the Golan Heights and Sinai Peninsula. In the aftermath of the ensuing short-lived war, which saw the attackers repelled prior to the imposition of a ceasefire on 25 October, the Organization of Arab Petroleum Exporting Countries (OAPEC) imposed an oil embargo against countries such as Canada that supported Israel. As a result of the embargo and a cutback in production, the price of oil increased by 70 percent that October, and a further 130 percent in December. The energy crisis that ensued would wreak havoc on the Canadian and American economies, which had been built on the availability of cheap oil. Although the embargo was eventually lifted in March 1974, the experience of government-imposed rationing and price controls led to a dramatic growth in interest in alternative energy sources and conservation in Canada and much of the industrialized world. As the level of funding available in this area began to increase, Pollution Probe was ideally positioned to capitalize, with important ramifications for its future structure and operations.

Recycling/3Rs Team

Like the Energy and Resources Team, the Recycling Team was decidedly policy-driven. Headed by Gregory Bryce, its major emphasis was on conveying the importance of recycling within a province-based waste program. Concern for dwindling resources and a growing waste problem led Environment Minister James Auld to announce the formation of the Solid Waste Task Force in the autumn of 1972.[23] The task force's terms were sufficiently broad, including "any aspects associated with the production, handling, and reclamation or disposal of refuse," and its primary objective was "to ensure that deleterious effects on the environment are minimized, and that socio-economic factors are given consideration."[24] Pollution Probe was invited to provide a representative to the task force, and announced in the November 1972 edition of its newsletter that it was "encouraged by the terms of reference," although it was concerned that the twelve-person body was dominated by industry groups.[25]

Given the full-time demands of serving on the task force, including its subsidiary Beverage Packaging and Milk Packaging working groups, the decision was made to hire a new staff member, Peter Love, to fill this role. In hiring Love, the organization turned to a familiar face. A volunteer at Pollution Probe from the time of its founding through his graduation in 1971, Love joined sister-in-law Ann as the first of many family members to work for the organization, while his father Gage was a long-time member of its Board of Advisors.[26] Peter Love's familiarity with the organization's operations served him well, as he was hired just weeks before the provincial task force's inaugural meeting on 2 November 1972. In the meantime, he composed Pollution Probe's preliminary submission, which advanced the lofty goal of seeing "Ontario advance along the path towards *zero garbage,* obtained through the absolute minimizing of throughput combined with the recycling of all the rest of the waste."

Love's submission introduced one of Pollution Probe's most important contributions, the waste hierarchy, to the ongoing garbage discourse. According to this hierarchy, the province's first priority should be to "reduce throughput," which could be accomplished by educating consumers, creating a Consumer Product Review Board that would eliminate overpackaging, and encouraging the "reuse of materials" by banning non-refillable containers, increasing deposits on returnables, and standardizing containers "to promote easier handling." The second priority was to promote "recycling as an alternative far superior to burning and/or burying," which could be accomplished by having the province institute a preferential purchasing policy for recycled materials, making recycling equipment tax-exempt, taxing non-recyclable items, and taxing items that were manufactured from virgin resources, while at the same time eliminating benefits such as subsidized freight rates, depletion grants, and tax concessions enjoyed by resource extraction industries.[27] As he later explained, "there was a priority in what needed to be done. We should reduce as much as possible of this garbage. We should reuse as much [as possible] after that. And then third, and last, we should recycle. At the very end there would be so little waste that we wouldn't have to worry about it."[28] By January 1973, Pollution Probe was promoting the waste hierarchy as "the 3Rs," which has since grown synonymous with the recycling movement. Shortly thereafter the Recycling Team was rechristened the 3Rs Team. It is noteworthy that the 3Rs originally referred to "Reject, Re-use, Recycle." However, Pollution Probe replaced "reject" with the more familiar "reduce," as the former term was deemed to be too harsh.[29]

On 19 December 1974, the Solid Waste Task Force's report was tabled in the provincial legislature. Its chief recommendations were that the Ministry of the Environment "actively pursue a comprehensive waste management policy aimed at reducing and recycling as much of Ontario's solid waste as possible," including incentives for "industry to research and develop waste management technique and markets," and that it should create "a permanent body, i.e., the Solid Waste Management Advisory Board, to investigate and advise him [the Minister of the Environment] on all aspects of waste management policy." Drawing from the task force's own experience, the report suggested that this advisory board be an independent body drawn from interested parties outside the Ontario civil service and the industries immediately affected.[30]

Pollution Probe was clearly disappointed with the final report. Love argued that it "is filled with meaningful data not reflected in its recommendations." Conspicuously absent from the latter was a ban on non-refillable beverage containers, without which Pollution Probe argued the province could not begin to adequately address its solid waste problem. As Love pointed out, "as long as non-refillables exist they will eventually become garbage, and we will have done little to reduce the growing solid waste problem." Still, Pollution Probe found reason for optimism in the recommendation that a permanent Solid Waste Management Advisory Board be established to advise the Minister of the Environment. As Love wrote: "If we had an advisory board 15 years ago, non-refillables would not be the problem that they are today." Qualifying his optimism, however, he noted that such a committee could succeed only if the government chose its members carefully.[31] In 1975, the Ministry of the Environment established the Ontario Waste Management Advisory Board, which "sought to foster and promote comprehensive government policies and programmes to conserve resources, reduce and recycle waste, and eliminate harmful waste effects."[32] However, the board was staffed by government mandarins and was dismissed as ineffective by Pollution Probe on release of its first report in March 1976.[33]

As the 3Rs Team continued its research into existing recycling systems, it grew increasingly bullish on a technology offered by the Black-Clawson Company of Franklin, Ohio. Whereas the standard recycling program required participants to sort their materials prior to collection, the Black-Clawson plant was a fully mechanized system. The hallmark of streamlined operations, unsorted recyclable materials were loaded on a conveyor belt at one end of the recycling plant, with metals passing through unshredded while other materials were then broken into smaller pieces, filtered, and sorted by colour and texture. Pollution Probe backed this system for two reasons. First, it offered the most user-friendly approach to recycling for the general public. Rather than having to learn to separate their recyclable waste into various categories, the system limited the options available to garbage and recyclables. Second, Pollution Probe felt that merely requiring citizens to divide their waste into two streams would enable them to avoid the stigma of handling "garbage," a perceived impediment to involvement for some.[34] In March 1973, five members of Pollution Probe visited the Franklin plant, which handled fifty tons a day. Sufficiently impressed with what they saw, Gregory Bryce noted in a *Globe and Mail* opinion piece: "We're convinced that mechanical recycling plants are a solution in part to Toronto's huge solid waste problem – right now."[35]

Pollution Probe's early efforts to cajole the Toronto and Ontario governments into funding a mechanized plant were roundly rejected. James Auld wrote on 25 May 1972 that he had no intention of spending public money on such a project "until markets for reclaimed products are made attractive."[36] While the market for reclaimed newspaper, long the cash cow of recycling, collapsed in 1974 as a result of the market's failure to keep pace with growing demand, increased difficulties in acquiring new landfill sites, combined with industry's push to prevent additional packaging restrictions, led the provincial government to announce plans in October 1974 for a provincewide system of mechanized recycling plants. The first six plants, their $17 million total cost to be shared with the municipalities, would be built in London, Sudbury, Kingston, and Metro Toronto, which would receive three. Environment Minister William Newman envisioned the expansion of this system into every major Ontario community, at a cost of $500 million, over the next fifteen years.[37] The 3Rs Team, which had spent much of 1973 and 1974 working on reports demonstrating the feasibility of recycling, were encouraged by the government's announcement. Proclaiming in autumn 1975 that recycling was "close to being institutionalized," the 3Rs Team once again changed its name – this time to the Garbage Team – and shifted its focus to reducing throughput.[38]

Urban Team

The Urban Team, led by Lynn Spink (who went by the name Marilyn Cox while working at Pollution Probe), presented an approach to environmental problems unique to Pollution Probe.[39] As Cox explained in a letter to the Community Planning Association of Canada, the team had been "created to deal with the very specialized problems of the city environment. The Team is devoted to a combination of study and action relating to the environmental implications of development and transportation, and the process by which decisions affecting the people who live in cities are made."[40] Operating in a manner akin to the community organizers with the Company of Young Canadians and the Toronto Community Union Project, the Urban Team aimed to empower those living in Toronto's downtown by serving as an advocate while also providing citizens with the resources and information necessary to combat their environmental problems.

One of the Urban Team's chief concerns was the process of development prevailing in Toronto. Toronto's 1969 city plan set aside large sections of the downtown core for high-density developments, in particular the area between Bloor and Queen, and west of Jarvis to Spadina. The city plan ushered in a new era of construction. As historical geographer James Lemon has noted, there were no apartment buildings taller than twenty storeys prior to 1965; by 1973 there were 142.[41] The Urban Team acknowledged that these buildings did have certain environmental benefits, as they required less land and were more efficient to heat, but argued that the developments failed to address the need for family dwellings in the city, created an apparent link between the high-rise lifestyle and mental and physical health issues, provided a visual "assault [on] our aesthetic sensibilities," and resulted in associated ecological problems, such as wind tunnels and the reduction of direct sunlight.[42] The Urban Team was particularly concerned that the developers often relied on underhanded tactics such as blockbusting – pressuring residents to sell their homes by purchasing the surrounding buildings and then allowing them to become rundown. One area of particular concentration was St. James Town, with fifteen high-rises containing close to six thousand apartment units. When city councillor John Sewell led a campaign to prevent an expansion of this project to the area immediately south of St. James Town, the Urban Team played a supporting role, serving as an information clearinghouse and a coordinating body for concerned residents and community groups. It began publishing the monthly newsletter *Whose City?* which was designed to keep community groups abreast of important meetings at City Hall as well as development-related news from throughout Toronto.⁴³

Another important activity of the Urban Team was keeping track of real estate purchases of the city's major developers. As Spink explains:

> At the time there were a large number of land assemblies going on secretly, by developers who wanted to redevelop areas of the city. McCaul Street, Quebec [Avenue], Gothic [Avenue], Beverly Street, south of St. James Town. We discovered that there was an organization used by the real estate industry called Teela Marketing, which published regularly a record of real estate transactions ... We got a subscription to the Teela Marketing survey and mapped for residents' groups and with residents' groups all the sales that we could identify where land assemblies were going.⁴⁴

Likening this to "a distant early warning system for residents," it removed the element of surprise from the land developers' arsenal and gave residents extra time to prepare a strategy.⁴⁵

The Urban Team's work on transportation issues was headed by Tom Murphy. Vocal critics of automobiles – in a 1973 speech, Murphy urged the Canadian government to phase out production of internal combustion engines by 1980 – the team welcomed the provincial government's abandonment of plans for the Spadina Expressway but doubted that its planned replacement, the Spadina Rapid Transit System, was much of an improvement. Its March 1972 submission to the Metro Transportation Committee explained that the seven routes then under consideration for the subway "all involve irreparable destruction to the ravines ... a considerable loss of trees" and the "destruction of one or more stable communities." Thus, although the brief acknowledged the necessity of a north-south rapid transit line, it urged those responsible for selecting its route to take "into account the very real environmental costs involved."⁴⁶ This position was reiterated in a brief to the Ontario Municipal Board in April 1973, in which Murphy attacked the favoured route, which would have resulted in the destruction of the Cedarvale and Nordheimer ravines.⁴⁷

By 1973 most of the Urban Team's time and energy was focused on the Dufferin-Davenport neighbourhood. A predominantly working-class community, it featured a number of factories, including Prestolite, which manufactured batteries; Universal Fur Dressers and Dyers, a tannery; Union Felt; Toronto Foundry; and Kelson Springs, which produced mattress springs. These factories were classified as "heavy industry" by the city but because they predated the 1952 zoning regulations, they were allowed to continue operating. Unsurprisingly, they generated considerable pollution. The team's aim was to coordinate the efforts of local residents' groups and help them obtain data from the factories and the Ministry of the Environment, an often daunting task in the days before access-to-information legislation. While Dufferin-Davenport contained many long-term English-speaking residents, it was also home to a substantial Italian immigrant population. In order to effectively reach the latter, an Italian speaker, Connie Procopio, was hired. Procopio, Cox, and Irene Harris, who had previously organized local residents' groups, served as the core of the project staff.[48]

The Urban Team continued operating until September 1974. The stated aim of the Dufferin-Davenport project had been to empower members of the community to effectively address their own concerns. The October 1974 edition of the *Probe Bulletin* explained that "they still have their problems, but now the community is dealing with them itself: when problems arise in the future, they will know how to go about finding the right answers."[49] The dissolution of the Urban Team coincided with a career shift for Spink, who was offered a job with the city as a neighbourhood planner. Given her lack of any formal training in city planning, it was apparent to Spink that the offer came as a result of her work with Pollution Probe. She accepted and began mapping out a plan for the King-Spadina area.[50]

As noted at the outset of this chapter, the work of the Urban Team raised eyebrows among those who questioned the relevance of some of its projects. This sentiment was echoed within Pollution Probe itself, as some members looked askance at the team. As Gregory Bryce pointed out, some team members, most notably Murphy, had a sophisticated class-based analysis of events that seemed particularly radical, and "a lot of what they got involved in, I think some of us had a bit of trouble seeing what the connection was [to environmental issues]."[51] Peter Middleton recalls: "I had to convince Don Chant that this [creating the team] was okay, and I had to convince some donors or fundraisers that this was okay because it was in many ways the most radical thing we were doing."[52] Philosophical clashes were inevitable, primarily with the Energy and Resources Team. The tension, clearly evident during Pollution Probe's meetings, was rooted in differences between the team leaders' approaches to problem solving. Spink's Urban Team believed in grassroots mobilization and gave priority to consultation with affected communities. The Kelly-led Energy and Resources Team, meanwhile, concentrated on getting its information, backed by scientific data, to the corridors of power, an approach in part informed by the Energy and Resources Project's failure to gain traction with the public in its 1970 lever campaign. A unique entity within Pollution Probe, the Urban Team was in many ways ahead of its time, foreshadowing the environmental justice movement that first gained prominence in the United States during the early 1980s.[53]

ACTION TEAM

Pollution Probe had cut its teeth organizing high-profile events such as the Don River funeral and the public inquiries into dead ducks and air pollution, but by 1972 its work was increasingly taking place behind the scenes and was policy-driven. Although this was indicative of a group in the process of institutionalization, executive director Peter Middleton believed that its ability to effectively orchestrate public action campaigns was central to its identity. He reasoned that Pollution Probe's willingness to undertake such initiatives imparted a certain mystique to the ENGO, kept its opponents off-balance and unable to predict its actions, and helped it maintain its media presence. This resulted in the creation of the Action Team. Led by Ann Rounthwaite, Pollution Probe's former communications coordinator, the team was designed to identify egregious cases of environmental abuse and then, in cooperation with concerned locals, utilize Pollution Probe's media connections and organizational know-how to address the problem. In the process, Pollution Probe would resolve an environmental problem, create a new group of community allies, and add to its reputation as an effective operation.[54]

A situation tailor-made for the Action Team soon came to Pollution Probe's attention. Residents of the Borough of York had been complaining for over three decades that the Canadian Gypsum factory was seriously degrading their quality of life. The factory, which produced rock wool for housing insulation, was well known to authorities for its emissions of sulphur dioxide, which produced a noxious odour, as well as solid particulates ranging from dust to chunks three inches in diameter, which covered neighbouring properties. The Action Team established a working relationship with the Upper Humber Clean Air Committee, an ad hoc organization of concerned locals. It then set about producing a report detailing the long history of complaints, a correspondence log highlighting the run-around the Action Team received when seeking information from Canadian Gypsum and the provincial government, and emissions data showing that the plant was exceeding permissible limits, which they calculated cost the surrounding community $482,800 annually. The report also examined Canadian Gypsum's corporate composition, revealing the names and backgrounds of its board of directors and identifying its lawyers, the Toronto-based McCarthy and McCarthy, and its bankers, the Toronto-Dominion Bank – both of which had representatives on the board.[55]

On 13 March 1972, after receiving an advance copy of Pollution Probe's damning report, York Council delegated Mayor Philip White and the Board of Control to meet with James Auld, the provincial environment minister, and requested a review of the complaints. The following week saw two high-profile meetings on the matter. Mayor White, the York Borough Board of Health, the Ministry of the Environment's Air Management Branch, Pollution Probe, and the Upper Humber Clean Air Committee met with spokesmen from Canadian Gypsum. The latter emphasized that the factory operated in compliance with regulations – despite being fined twice the previous year for violating the Air Pollution Control Act – but the others were not swayed. The mayor demanded a clear timeframe for improvements to the factory's emission controls, while Larry Green of Pollution Probe's Action Team called for its closure.[56] The following day, the mayor met with Auld and presented him with a box full of fibre emissions from the factory's smokestacks as well as a petition signed by two thousand locals demanding government action against Canadian Gypsum. Auld acknowledged the complaints but rejected calls to shut down the factory, explaining "that under EPA [Environmental Protection Act] a stop order can only be issued when there is immediate danger to human life, health, or property. I am advised by my legal officers that we probably don't have enough evidence to establish this is the fact." He did note that his department had been pressing Canadian Gypsum on the issue, and that representatives from its parent company had scheduled a meeting with him the following week.[57] The following month, Canadian Gypsum announced that it would spend $645,000 in pollution abatement equipment, and that it would halt production should pollution exceed acceptable limits.[58]

Proclaiming victory in the affair, the Action Team produced a booklet detailing how it "'persuaded' the Canadian Gypsum Company to announce a thorough clean-up of its rock wool plant."[59] It was hoped that the booklet would demonstrate how to undertake similar campaigns, and the May 1972 *Probe Newsletter* noted that "the team is looking around for another major polluter to tackle."[60] According to Peter Middleton, however, it became increasingly difficult to identify "outstanding single point pollution horrors" in Metro Toronto.[61] The Action Team was therefore dissolved, with the understanding that the remaining teams would launch similar campaigns when warranted.

Land Use Team

Even as the Action Team disappeared, 1973 saw the creation of two new teams. The Land Use Team addressed an area of concern that Pollution Probe had raised as early as 1971. The team's premise was that a clear relationship existed between land-use policy and environmental problems. Rather than waiting to treat the environmental symptoms of poor planning, the team would anticipate problems relating to traffic congestion, waste treatment, and air pollution, and prescribe solutions. Pollution Probe's interest in land use was spurred by the provincial government's 1970 announcement of the Toronto-Centred Region Plan, which aimed to shift the emphasis of development in the province from the western edge of Metro to the eastern edge, while also curtailing growth to the north. This led the federal government to announce, in March 1972, its purchase of land in Pickering Township for a new international airport; the province also acquired land for a planned community to the south of this site.[62]

In April 1974, the Land Use Team submitted a brief to the Pickering Airport Inquiry. The brief highlighted concerns related to noise, air, and water pollution, as well as the destruction of agricultural and recreational land, but the major focus was on the availability of Canadian oil. Given the recent OAPEC embargo, the brief argued that oil might not be readily available within three years and could cost considerably more, and that these factors may make air travel less popular, rendering the airport unnecessary. Rather than investing in energy-intensive travel systems, the Land Use Team maintained, the government should invest in high-speed rail.[63] The following month, the team released *The Tail of the Elephant: A Guide to Regional Planning and Development in Southern Ontario*. This handbook, designed to spark interest in the subject among the general public, provided readers with an overview of regional planning and its impact, and challenged them to consider whether "the growth that the government plans for Ontario [is] in the best interests of all the people of the province."[64] Pollution Probe claimed that the handbook sold well but expressed disappointment in October 1974 that it had failed to spark public debate on the issue. Noting that the province lacked a population that was well informed on land-use policy, the organization disbanded the Land Use Team. As explained in the *Probe Bulletin:* "It could be argued that Probe's job is ... to bring the whole planning issue to peoples' attention, and then coordinate action on it. That was exactly our intention, but we were before our time."[65]

Caravan Team

The Caravan Team was created out of a desire to keep Pollution Probe in the forefront of Ontarians' minds. Pollution Probe created a mobile multimedia show on environmental issues that travelled across the province with a forty-five-foot trailer. The project required extensive funding, costing $214,000 for the initial eighteen-month period.[66] Beginning early in 1973, the team travelled from town to town, giving presentations to schoolchildren, community groups, and government officials, and conducting interviews with local media. Described by Pollution Probe as "a clearinghouse dispensing information, advice and encouragement to people who wanted to do something about local environmental problems,"[67] its staff of five, headed by Peter McAskile, gave eight hundred presentations in 1973. According to Joe Warwick, Pollution Probe's media coordinator from 1972 through 1973, the Caravan Team had little difficulty attracting attention from the local press, particularly small-town newspapers, which gave the project prominent coverage. The team was brought back for a second tour in 1974.[68]

Conclusion

The financial boon that fuelled Pollution Probe's growth came to a halt in October 1973 when the energy crisis struck, leading to a recession. Government make-work initiatives such as Opportunities for Youth and the Local Initiatives Program were severely curtailed at first and then cancelled outright in 1977, and corporations and foundations were forced into a new era of austerity. The recession did not immediately affect Pollution Probe's bottom line: during the fiscal year ending 30 September 1974, it raised $329,097, which had been secured before major budget cutbacks were implemented. By the outset of 1974, however, it had become apparent that Pollution Probe could not sustain its operations at the same level as before. This point was made abundantly clear by Peter Middleton at a staff retreat in January and re-emphasized in May.[69] When the cuts finally came in the fall, the payroll was reduced from a high of twenty-five staff in October 1973 to just thirteen. As was explained at the time, "in many ways it's a more manageable number, but it means fewer projects if we are to avoid spreading ourselves too thinly."[70] Forced to prioritize, the organization dissolved its Urban and Land Use Teams while renewing its emphasis on education, energy, and waste issues.

The economic reality of working at an ENGO now began to assert itself. In the summer of 1974, Peter Middleton left the organization. Pushing thirty – a ripe old age in Pollution Probe – he parlayed his experience and contacts into a for-profit consulting firm, Peter Middleton and Associates. Specializing in energy issues, the firm quickly secured a major contract from the federal Department of Energy, Mines and Resources (EMR). Middleton would also play an important role in the creation of the Institute of Man and Resources on Prince Edward Island. Brian Kelly left Pollution Probe at this time as well. Recently married and intent on raising a family – a far-fetched notion on Pollution Probe's minuscule salary – he accepted a position in Ottawa with the Office of Energy Conservation, a newly created branch of the EMR.[71] As Peter Love, who soon joined Middleton as his first hire, points out, "being at [Pollution] Probe was never a career. It's not as if anybody was looking at a corner office saying 'In two years I'll be here, and then I can be here.'" Staffed by twenty-somethings, the work paid poorly and demanded strenuous hours, with fourteen-hour days being commonplace. "It was a burn-out job," recalls Love, "but you loved it. It was what needed to be done so you did it."[72]

5

The Changing Environmental Landscape

THE ENVIRONMENTAL MOVEMENT emerged during a time of economic prosperity. In the wake of the 1973 economic downturn, much of the government and public enthusiasm for environmental initiatives was redirected towards the economy. Whereas politicians once felt obliged to publicly align with the movement, by 1977 many politicians in Canada and the United States had grown, in the words of Robert Paehlke, "openly disdainful of environmental activism."[1]

The movement's declining profile was exemplified in the 19 May 1977 Toronto *Star* article "Pollution Probe's Alive and Fighting for Environment." Written in "Where Are They Now?" style, the article's opening lines highlighted the one-time media darling's declining public status. "Remember Pollution Probe, that group that operated out of the University of Toronto in the early 1970s, the group that wanted to change the world by making it cleaner? Probe is still around, and the name of the game is change, but they're playing it differently now." "Sure, we're still here, because the problem is still here," staff member Jo Opperman is quoted as saying. "We're still trying to change things, but we've changed the way we play the game."[2] As Opperman intimated, Pollution Probe no longer engaged in the high-profile action campaigns of its early days, having shifted its attention to behind-the-scenes policy work. Just four months later, a similar feature ran in the University of Toronto's *Varsity*. Opening with the question "Whatever happened to the public concern over pollution of a few years ago?" it notes that the issue had been supplanted by economic concerns. "Although the attention of the public has shifted," the article continued, "Pollution Probe still lives, and despite the lower profile of this independent public interest group, it still remains concerned about pollution."[3] Pollution Probe lived on, but it seems few took notice.

The years 1975 to 1980 were a transitional period for Pollution Probe, and for the environmental community more broadly. As Pollution Probe struggled to define its role within the broader movement, its offshoot handling energy issues – the newly formed Energy Probe – successfully carved out a niche. Lingering financial difficulties, however, would lead Energy Probe to sever all ties with the Pollution Probe Foundation. Meanwhile, the local environmental community continued to be refashioned as Greenpeace, with its direct action tactics, established a presence in Toronto, as did the business-minded Is Five Foundation, which would emerge as the most influential group addressing the recycling issue.

The Birth of Energy Probe

In January 1975, the Pollution Probe Foundation underwent its most substantial overhaul, scrapping the team format and replacing it with two semi-independent partner projects, Pollution Probe and Energy Probe. This transformation was a direct result of the Organization of Arab Petroleum Exporting Countries oil embargo and the heightened public and government interest in alternative energy and conservation. Given its established interest and expertise in this area, the Energy and Resources Team was uniquely positioned to contribute to this discussion but it felt confined by the Pollution Probe moniker. Coordinator Brian Kelly explained in a December 1973 letter: "For some time now ... we have felt the need to speak out on energy issues from a broader basis than our environmental perspective; to consider social and economic questions such as cost, control, development and other aspects affecting the public interest." To accomplish this, he raised the idea of creating a public interest group called Energy Probe. The proposed group would have a measure of autonomy from Pollution Probe, yet would maintain an affiliation with the Pollution Probe Foundation, a body formed for legal and fundraising purposes, in order to continue utilizing the infrastructure and assorted benefits enjoyed by the Energy and Resources Team.[4]

A separate identity was also central to the team's ability to fund its activities. More focused on national public policy matters than the other Pollution Probe teams, it also required more financial support, but despite greater fundraising potential for the sort of activities it was undertaking, the Pollution Probe name gave some sponsors pause. Sanford Osler, Energy Probe's first leader, recalls that "some people were more interested in energy issues than pollution issues in terms of sponsors, because to some extent the spotlight changed in '74 [due to the energy crisis] from pollution issues to energy issues. We didn't want to go on our own fundraising-wise – we still wanted to be part of the Pollution Probe Foundation – but we felt the name change would appeal to certain sponsors."[5]

In January 1974, a Pollution Probe committee was established to study the proposed organizational change. Its report provided a series of wide-ranging suggestions. The committee recommended "that greater freedom be given teams to spend money when and as they see fit" and "that teams be given increased autonomy in setting goals, defining issues and developing strategies." The report also recommended "that the name of the Energy and Resources Team be changed to Energy Probe in an attempt to solve some of the problems outlined in the team brief concerning external relations." The report specified that money raised by Energy Probe would be used to meet its operating expenses, but that Energy Probe would continue to have access to the general funds used for existing Pollution Probe projects. The report stopped short of suggesting that the new group be given fully independent powers, noting that: "We all depend upon the tradition and public credibility of the Pollution Probe name for our legitimacy and our rights to solicit money and take action. We have a duty to manage the money given to us responsibly, and consequently, we could not accept the removal of final accountability to Pollution Probe for Energy Probe's action."[6] These suggestions were adopted and Energy Probe was elevated to a project of equal status with Pollution Probe within the institutional home of the Pollution Probe Foundation.

Energy Probe held its official launch on 16 January 1975, explaining, in a prepared statement, that its "objective is to stabilize average per capita energy consumption in Ontario, and in Canada as a whole."[7] The original plan was for Brian Kelly to lead Energy Probe, but when he left the organization in late 1974, the reins were handed to Osler, who had worked with Kelly on energy issues since the summer of 1971. In the summer of 1975, Energy Probe's leadership changed once again when Osler decided to return to school. His decision was prompted by the changing awareness of energy issues within Canadian society. The Energy and Resources Team had long focused its efforts on trying to convince the public that energy problems were both real and serious. In the aftermath of the 1973 energy crisis, Canadians became increasingly interested in potential solutions. Osler felt that he was not adequately prepared to provide answers, particularly with respect to economic factors, and this led him to pursue a master's degree under natural resource economist Dr. John F. Helliwell at the University of British Columbia.[8] After completing his degree, he returned to Toronto in 1977 and worked to incorporate his ideas at Ontario Hydro.

Thus, in less than a year Energy Probe lost the two key figures who had guided it since its earliest days as the Energy and Resources Team. Without them, the group began forging a new identity. Gone was the idea of utilizing Canada's resources in a bid to check US population growth and consumption. Instead, the group developed a strong anti-nuclear focus. Osler says of his and Kelly's approach: "I don't think we took a firm stand against nuclear. We were sort of nuclear watchdogs, but I don't think we were anti-nuclear."[9] To this effect, the Energy and Resources Team had limited its critiques of nuclear energy on the grounds that it detracted from the message of conservation and efficiency. In this respect, the team's attitude was quite typical of the Canadian environmental movement. While there had been an anti-nuclear movement in Canada dating back to the 1959 formation of the Canadian Campaign for Nuclear Disarmament and the Combined Universities Campaign for Nuclear Disarmament, these groups focused on the military application and the expected radioactive fallout from the use of nuclear weapons. Canadian activists did not turn their attention to the country's domestic nuclear power program until 1974, when India detonated a nuclear bomb utilizing plutonium manufactured in a reactor sold by Atomic Energy of Canada Ltd.[10] The change in attitude is best symbolized by Energy Probe's hiring of Barry Spinner in 1975. Spinner obtained a degree in chemical engineering from the University of Toronto in 1969 and returned in 1973 to pursue a master's in engineering that focused on nuclear chemistry and nuclear engineering. While working on his thesis, he met Syed Naqvi, a political refugee who had been an attaché at the Pakistani embassy in Paris. Naqvi explained to a dumbfounded Spinner that Pakistan had recently purchased technology from France that enabled it to develop nuclear weapons. The revelation had a profound impact on Spinner: "He told me all this and I had a serious crisis of ... technological faith." Feeling that his graduate work was merely preparing him to abet nuclear proliferation, he left the program and joined Energy Probe as its nuclear specialist. Spinner hoped that by raising awareness of nuclear issues among Canadians, he could change public opinion, stop Atomic Energy of Canada from selling to Third World countries, and help turn the tide against global nuclear proliferation.[11]

Ontario Hydro, meanwhile, was in the midst of an ambitious nuclear expansion plan. Responsible for supplying the province's growing energy needs, which it estimated required a 7 percent annual increase in capacity, Ontario Hydro opened its first nuclear power plant at Douglas Point in 1968, with a second plant, the Pickering Nuclear Generating Station, already under construction. In 1974 Ontario Hydro announced plans to build an additional twenty nuclear plants over the next quarter-century. These plans ground to a halt in 1975 when, facing cost overruns, the public utility announced its intention to institute a 27 percent rate hike.[12] The province responded by appointing Dr. Arthur Porter to chair the Royal Commission on Electric Power Planning, which was mandated to examine "the long-range electric power planning concepts of Ontario Hydro."[13] The Porter Commission, which supplied Energy Probe with funding to prepare its intervention, became a focal point for the group, providing it with a ready-made forum for critiquing the nuclear industry. Energy Probe maintained a twofold focus in its briefs. It argued that Ontario Hydro's plan for 7 percent annual growth, which would require capital expenditures of $80 billion by 1993, was both overblown and economically unviable. It was also argued that Ontario Hydro should abandon its focus on nuclear energy in favour of renewable energy and conservation.[14]

Energy Probe also engaged in public outreach on nuclear power. In December 1975, it released the handbook *CANDU: An Analysis of the Canadian Nuclear Program*. Noting that continuing "down the path towards an expensive energy-intensive nuclear, centralized, electric society" would prevent Canadians from pursuing "a lower energy, more decentralized, softer-technology society based upon conservation and renewable resources,"[15] the handbook highlighted a series of technical concerns with the CANDU (short for CANada Deuterium Uranium) program, particularly with respect to radioactive waste management, probabilities of accidents related to reprocessing technology, occupational health hazards, and security of the fuel cycle.

Opposing Trajectories within the Pollution Probe Foundation

Energy Probe expanded quickly. In little more than a year, its staff grew from three to seven. However, this growth belied major problems developing within the Pollution Probe Foundation. Although funding was available for work on energy-related issues, few donors were interested in Pollution Probe's environmental work. Between the fiscal years ending 30 September 1974 and 1975, its revenue fell from $329,097 to $119,128, bouncing back only slightly the following year.[16] The organization began having difficulty meeting payroll, which dampened the previously buoyant staff morale.

Shrinking revenue coincided with the departure of the last of the organization's old guard. In May 1975, financial czar Tony Barrett, the spark plug behind many Pollution Probe activities, left to pursue entrepreneurial opportunities in the field of environmental technology. One year later, Monte Hummel, who had succeeded Middleton as executive director, followed suit.[17] While the original intent had been to find a replacement for Hummel, the decision was eventually made to revert to a model by which decisions would be made collectively by Pollution Probe Foundation staff. In retrospect, this decision appears to have been ill conceived. In the midst of an economic downturn, the organization now lacked an identifiable leader who could serve as a mediator and at the same time be counted on to make unpopular but necessary decisions regarding operations.

Pollution Probe assumed a decidedly lower profile in 1975 and 1976. In the spring of 1975, it joined the Metropolitan Toronto Airport Review Committee (MTARC), a coalition of sixteen environmentally oriented groups that opposed the construction of the Pickering international airport, which had been given the green light in February. Working in concert with the pre-existing People or Planes, an organization formed by residents directly affected by the proposed development, MTARC worked to demonstrate that opposition to the airport was not merely a "not-in-my-backyard" issue. As noted earlier, opposition to the development was based on a belief that transportation planning should emphasize high-speed rail service, not oil-intensive airplanes, and on the fact that the airport was situated on prime agricultural land, an increasingly rare commodity in Metro Toronto. While People or Planes worked to keep opposition to the airport in the media – for example, occupying a house within the construction zone that was scheduled for demolition – MTARC lobbied policy makers. This two-pronged effort, coupled with the provincial government's abandonment of the Toronto-Centred Region Plan and a well-timed provincial election that saw the governing Conservatives fall to minority status, resulted in the airport plan's cancellation that autumn.[18] Pollution Probe also served as the coordinating secretariat of the Garbage Coalition, a federation of sixty-one anti-waste groups located throughout Ontario. An alliance of recycling advocates and localized groups opposed to the opening of landfill sites in their communities, it played a central role in lobbying the provincial government to abandon plans to open dumps in Hope Township and Pickering.[19]

While the Pollution Probe staff members interviewed for the 19 May profile in the Toronto *Star* suggested that its lower profile was part and parcel of the ENGO's new approach, those involved were clearly not content. As such, the ENGO attempted to increase its public profile throughout 1977 by orchestrating a series of short-term action campaigns. In April 1977, it launched the Pop Posse, which encouraged Ontarians to report violations of the Environmental Protection Act, particularly with respect to a recent amendment that required retailers to display a stock of returnable soft drink containers equal to, or exceeding, that of non-returnable containers. This was followed in September by the Boomerang campaign, which was designed to draw attention to the 975 pounds each person generated in packaging annually. This campaign encouraged consumers to voice their opposition to the waste of energy and resources on excessive packaging by mailing it back to the manufacturers. Pollution Probe also experimented with announcing awards, such as the Disposamaniac Award for companies found to be particularly wasteful in their packaging, and the Imagineering Award that recognized positive environmental actions undertaken by government and the business community. Although described as "action campaigns," they bore little resemblance to Pollution Probe's earlier efforts. Whereas the ENGO once strove to set an example by tackling major environmental problems head-on, now it placed the burden on concerned members of the public. Unfortunately, these efforts failed to make much of an impression on the media and the general public. As of January 1978 the Pop Posse, for example, had resulted in just one fine being levied, against Tamblyn Drugmart.[20]

The Pollution Probe Foundation also attempted to raise its profile through a bimonthly news magazine, the *Probe Post*, which was launched in 1978 under founding editor Robert Gibson. Gibson had a master's degree in political science but had no previous journalism experience and needed a crash course on the art of editing with the *Globe and Mail*'s Ross Howard.[21] Featuring material written by staff and volunteers with Pollution Probe and Energy Probe, the magazine highlighted projects underway at the Pollution Probe Foundation as well as issues of concern nationwide. The *Probe Post* proved to be a modest success and continued publishing until 1991.

The Pollution Probe Foundation's major undertaking during the late 1970s was the development of Ecology House. Initially a project of Energy Probe, the plan was to acquire an existing property and renovate it to highlight the manifold practical conservation technologies and alternative energy sources, such as solar energy, that middle-class urbanites could adopt in their households. The project was first announced in the foundation's 1975-76 annual report, and the first year saw its originator, Energy Probe's Richard Fine, exploring the technological applications and funding avenues for what promised to be a costly endeavour. In 1977, a three-storey Victorian building located at 12 Madison Avenue, in the heart of downtown Toronto, was acquired and renovations began. The city-owned property had been left in disarray after its use by a construction crew working on the Bloor Street subway line. Subsequently, the federal government purchased the building at a reduced rate and leased it to the Pollution Probe Foundation for five years at a dollar a year.[22]

Housing designed to demonstrate energy efficiency and renewable sources already existed, but Energy Probe's project was unique in two respects. Whereas other demonstration projects were typically located in rural settings, Ecology House was unmistakably urban. In addition, other projects tended to be built utilizing the most advanced technologies, regardless of price, but whereas these futuristic projects aimed to highlight potential achievements, Energy Probe's was a retrofit project, designed to showcase practical ways to save energy and money. In the end, however, Ecology House turned out to be a pricey undertaking and was possible only through funding from over thirty corporations and foundations (including Shell Oil, Dow Chemical, and the Bronfman Foundation), support from the municipal, provincial, and federal governments, and the labour of a phalanx of eighty volunteers. The renovations, which began the last week of June 1979 under the supervision of Brian Marshall, included attaching solar panels, super-insulating the building, replacing the roof, upgrading the wiring and plumbing, recycling water used in daily activities for use in irrigation, and installing a composting toilet. Renovations were completed in April 1980 – an event marked by an all-night party for the project's supporters – but fine-tuning delayed the official opening until October. The building became a popular destination for school field trips as well as workshops on topics such as passive solar heating and the utilization of alternative energy in buildings.[23]

Lawrence Solomon and *The Conserver Solution*

In 1978 the Science Council of Canada issued its landmark report *Canada as a Conserver Society: An Agenda for Action*. The main concept, first raised by the Science Council in a 1973 report, referred to a society that "promotes economy of design of all systems, i.e., 'doing more with less'; favours re-use or recycling and, wherever possible, reduction at source; and questions the ever-growing per capita demand for consumer goods, artificially encouraged by modern marketing techniques."[24] Eschewing discussion of socio-political matters in favour of more practical, technological possibilities, the Science Council, according to John B. Robinson and D. Scott Slocombe, was able "to argue that significant improvements in emissions reduction, land-use and resource-development practices, environmental protection, and the efficiency of resource and materials use were all possible through improved technological development without significant reductions in material standards of living."[25]

While the concept of the "conserver society" generated considerable interest in certain academic and environmentally inclined circles, most Canadians had a limited understanding, if any, of its meaning. Lawrence Solomon, a Romanian-born journalist, read the report and saw an opportunity to write a popular account for the general public. As he recalls: "I approached Energy Probe and Pollution Probe at the time to see if I could collaborate with them in producing that book. I thought that having them as a resource would help me in writing my book."[26] Solomon had already secured Canada Council for the Arts funding, and the Pollution Probe Foundation decided to endorse the project. Published in 1978, *The Conserver Solution* was jointly released in Canada and the United States by Doubleday. As Solomon explained in the introduction:

> Conserver principles only reconcile our environment with our economy; our ends with our means ... We have the capability today to begin phasing out all non-renewable forms of energy, such as gas and oil, and uranium, and begin phasing in a 100 percent renewable energy base, one founded on energy sources that will never run out on us ... We can begin phasing out our near-total dependence on continually depleting natural resources and begin phasing in a 100 percent recyclable economy, where our used resources are diverted from the dump and recycled for society's use. And we can strive for ever-increasing efficiencies, for doing more with less, for starting in earnest to unleash the imponderable potentials in the human mind, to produce an environmentally safe and economically sound place we'll be proud to pass on to our children. But we have to start now, or our room to maneuver will soon close in on us.[27]

The book received lavish praise from the likes of Maurice Strong, the former executive director of the United Nations Environmental Program, who proclaimed: "This book demonstrates convincingly that a Conserver Society is not only feasible; it can be an attractive, dynamic and exciting alternative to the gloomy future which the doomsters predict for us."[28]

Although the Pollution Probe Foundation's support was featured prominently on *The Conserver Solution*'s dust jacket and title page, the endorsement became a matter of contention after Solomon completed his initial draft. According to Chris Conway, an Energy Probe staff member: "It's creative, it's insightful, it's funny. It's a lot of really good things, but it didn't present the themes and the issues the way at the time a lot of people thought Pollution Probe wanted to present its public face. It's a little too much of a polemic, a little too casual with the facts."[29] The major problem with the book, according to critics within the Pollution Probe Foundation, was that it exhibited a wholehearted faith in the ability of the free market to self-correct problems related to the economy, the environment, and society. In some ways, this was not far from the organization's earlier ideas. For example, Chapter 7, "Paying Our Way," called for an end to hidden subsidies, arguing that industry should be charged the cost of any pollution incurred, rather than having taxpayers foot the bill. This idea was consistent with the ideas expressed in Pollution Probe's 1972 submission to Task Force Hydro, which argued that the cost of air pollution must be factored into Ontario Hydro rates, as well as the ENGO's earlier work on solid waste, which highlighted the fact that the true cost of recycling must take into account the savings from diverting material from landfill sites. Some of Solomon's ideas appeared to be overly ideological, however, with little apparent relationship to the environment. For example, Chapter 20, "Who Has to Do What," suggested a number of initiatives the government must pursue in order to achieve the desired outcome. Many of the ideas, such as "Adopt Total-Costing of Products," "Introduce Mandatory Life-Cycle Costing," "Remove Disincentives to Conserve," and "Promote Efficient and Durable Products," raised few eyebrows, but the suggestion to "Eliminate Red Tape by Simplifying Bureaucratic Requirements" did.

Even more alarming to some was the recommendation to "Eliminate the Minimum Wage and Social Welfare Programs." As Solomon explained: "The pricing mechanism of the free market is greatly distorted by the myriad of social welfare plans, and the minimum wage, which prevent people from working for nothing – if they choose – and companies from obtaining cheap labour where they can. The minimum wage has had questionable social value in Canada, since it is so low it only perpetuates the worker in poverty."[30] In place of the minimum wage and social welfare programs, Solomon proposed that a negative income tax system, set at the poverty line, be established. Family incomes over the poverty line would be taxed at a flat rate of 50 percent, while families earning less than the threshold would receive a negative income tax to bring them up to the minimum. These ideas were particularly controversial within the Pollution Probe Foundation. Chris Conway recalls that there was concern over the optics of endorsing a book that called for a pure free market while both Pollution Probe and Energy Probe continued to call for government intervention in environmental matters. There was also concern that Solomon never addressed the impracticalities of implementing his ideas. Although perhaps only Norm Rubin, Energy Probe's recently hired nuclear specialist, was comfortable with *The Conserver Solution* in its entirety, after a lengthy internal debate it was decided that they would release the book. The rationale was that the organization should be the conveyor of new ideas and approaches – "Plus, we were really, really tired of talking about the issue," states Conway.[31] Following the release of *The Conserver Solution*, Solomon returned to Energy Probe as a full-time volunteer, and also wrote a column for the *Probe Post*.

Energy Probe's Ottawa Office

Energy Probe had long considered the possibility of opening an Ottawa office. Given the federal government's jurisdiction over natural resources, a steady series of meetings were held in the city, dating back to the early days of the Energy and Resources Team. The office finally became a reality in 1977 when David Brooks joined the staff. Unlike most of his colleagues at the Pollution Probe Foundation, Brooks was not a young idealist on the threshold of his professional career. A native of Massachusetts, he held an MS in geology from the California Institute of Technology and a PhD in economics from the University of Colorado. He had moved to Canada in 1970 to become chief of the Mineral Economics Research Division of the Department of Energy, Mines and Resources (EMR), and in 1973 was named the founding director of the EMR's Office of Energy Conservation. A small operation – originally consisting of Brooks and his first hire, ex-Pollution Prober Brian Kelly – it was charged with developing a national energy conservation policy. After three and a half years, Brooks was ready to move on. He and his wife took a prolonged vacation in Europe and contemplated their future. Their children were grown and they were in a position to take financial risks, so he opted to pursue his research interests in the employ of Energy Probe, where the financial rewards were less certain but the potential for personal fulfillment was greater. He explained in an interview:

> I thought we'd had most of the fun times in the Office of Energy Conservation. It was inevitably going to be bureaucratized, which I don't say as a criticism. It's inevitable in government when you take a ginger group and then it begins to spin and eventually you've got to fold it back into the bureaucracy, and that was happening. But I also wanted to explore new areas ... We were just getting into the notions of soft energy paths and I really wanted to explore that more and see where we could go with a much more conservation[ist] program than the government would ever countenance.

Although Brooks felt Energy Probe was "by far the best group around," he was not willing to relocate to Toronto. Instead, he was assigned the role of Energy Probe's Ottawa liaison, with the requirement of attending at least one staff meeting in Toronto per month.[32]

Brooks's interest in soft energy paths – a move from capital-intensive, high-technology energy solutions towards sustainable technology and conservation – led him to author *Zero Energy Growth for Canada*. Written over the course of a year, the project was made possible by a Rockefeller grant secured by Lawrence Solomon. Published by McClelland and Stewart in 1981, the book linked the idea of zero energy growth to the conserver concept in Solomon's 1978 work. Brooks wrote:

> This book ... has been written as a contribution to the growing debate surrounding the idea of a conserver society – a society that depends less upon nonrenewable resources, material goods, and high technology and more upon renewable resources, human services, and appropriate technology ... Indeed, it is impossible to conceive of a conserver society in the absence of zero energy growth (perhaps even slightly negative growth) combined with reliance on dispersed renewable sources of energy. The alternatives are all ultimately either infeasible or undesirable.

The book also called for a paradigm shift, from the "current focus on the efficiency of energy use toward what might be called the *ethics* of energy use, *away from what energy can do for us and toward what we ought to do with energy.*"[33]

Greenpeace Toronto, Energy Probe, and the Nuclear Critique

In September 1971, Greenpeace revolutionized the role of environmental activists by bringing non-violent direct action to the movement. Whereas Pollution Probe earned its early reputation with its action campaigns, which saw it utilize the media to focus attention on environmental concerns, Greenpeace raised the bar by targeting objectionable activity with tactics such as occupations and sabotage. Its daring adventures captured the attention of many would-be environmentalists, leading to a quick expansion of affiliate groups. Unlike Pollution Probe, whose activities and media coverage were confined to Canada and inspired a plethora of affiliates across the country, Greenpeace addressed matters of global concern, was covered by the international media, and inspired the creation of affiliate groups worldwide. As Rex Weyler explains: "The affiliations remained informal, generally based on some individual having stepped forward and taken an interest."[34] Such was the case in Toronto, where John Bennett opened a Greenpeace office in autumn 1975. Employed by the University of Toronto Student Council as the secretary to the Executive Council, he had been involved in left-wing politics throughout his undergraduate studies but had no experience with environmental issues before attending an on-campus lecture delivered by Greenpeace co-founder Bob Hunter. For Bennett, "it felt like my kind of organization,"[35] and when he heard Hunter comment that the group could use a foothold in the city, he promptly offered his office space. After he left his campus job, the Greenpeace operations were based in his apartment. In January 1977, having received a Local Initiatives Program (LIP) grant for the "Greenpeace Toronto Education Project," the office moved to a storefront on Gerrard Street. This was followed by a succession of offices in downtown United Churches when the grant ran out.[36]

Dan McDermott, one of Greenpeace Toronto's first members, provides insight into the group's motivation. A native of Rochester, New York, he had been living in Toronto with his Canadian wife. After losing his job at a printing plant and finding himself with a decent severance package, he decided to dedicate his time to an environmental cause. There was little debate over which group he would affiliate with. As he explained: "I was a veteran of the sixties, and by the mid-seventies was noticing that it all seemed to have dissipated. And then along comes this organization which got rubber boats in between a harpoon and a whale, and immediately the approach captivated me." Pollution Probe was still the largest environmental group in Toronto, but McDermott was completely uninterested in joining it. In an indictment of its increased emphasis on behind-the-scenes work, he described the city's oldest ENGO as "wimpy" and "ineffective." "I wanted something that was more active, more cutting edge," he recalls, adding that "there was a certain glamour to being with Greenpeace in those days."[37] Ironically, just a few years earlier it was the allure of working with Pollution Probe that inspired many first-time environmentalists to take action.

The first year was rather blasé by Greenpeace standards, mainly consisting of selling pins to raise money, which was then sent to the organization's Vancouver office.[38] In the Toronto chapter's first brush with direct action, Dan McDermott participated in Greenpeace's second anti-sealing expedition, in March 1977. McDermott was content to play a supporting role to the Vancouver headquarters, but John Bennett saw things differently: "We were sitting in Toronto, and wanting to be involved in a direct action organization. The only action was being organized out of Vancouver ... It seemed to me we should be organizing our own things."[39]

Doug Saunders was a relative newcomer to Greenpeace Toronto in the summer of 1977. Having recently taken a leave from his PhD studies in photochemistry at the University of Toronto, he had grown particularly frustrated with the Porter Commission hearings on Ontario Hydro's long-term planning. The commission, he notes, "really implemented my understanding and recognition that our energy future was being decided by a small group of technical experts and a small group of political types who had decided that ... the future and viability of AECL [Atomic Energy of Canada Ltd.] was more important really than the health and well-being of Ontarians." Although he consulted with Energy Probe during the Porter Commission hearings, his own background in direct action – he had trained in New England in the conduct of non-violent opposition to nuclear station construction – led him to join the Toronto chapter of Greenpeace:

> I always appreciated and was drawn to some of the work Greenpeace had done because I felt it was important to capture people's hearts as well as their minds in terms of environmental issues and I felt that Greenpeace in particular saw its role as kind of being out front and drawing attention to issues that other groups could then follow up and provide some of the more well developed arguments to support it.[40]

Concerned that opponents of nuclear energy were being given short shrift by the government, the members of Greenpeace Toronto were itching for action. At this juncture, they were inspired by Tony McQuail, an outspoken nuclear opponent who lived near the Bruce Nuclear Generating Station, then under construction. McQuail, a Quaker farmer, had initially considered taking his team of horses to plough on the construction site, but this idea was dashed because "it wasn't good land, so I couldn't make the argument that they were wasting good land." Instead, he turned to the members of Greenpeace Toronto and suggested that they attempt to breach security at the station in order to highlight its susceptibility to a terrorist strike.[41]

The plan was simple, as Saunders explains: "In Greenpeace-style the idea was to canoe in early in the morning and to plant a banner on the containment building of the reactor, and to leave before anybody caught on to it, and then do the media work around that." The stunt, carried out on 11 July 1977, began well. At 4:00 a.m., Saunders, Bennett, and Rich Curry were dropped off by McQuail, and paddled their rented canoe across the bay to an area close to the nuclear facility. From a safe cover, they observed that security consisted of a lone patrolman circling the site every forty-five minutes. Rushing in and testing the doors to various buildings onsite – according to Bennett, the doors leading to the waste pool were unlocked – they approached the containment building and unfurled their Greenpeace banner. Saunders recalls with a laugh, "We probably needed to have a banner that was five times larger than what we had."[42] At a mere six by three feet, it was hardly the photogenic prop they desired. Efforts to capture the moment on camera were spoiled when Curry forgot to use the flash on his Kodak Instamatic, while Saunders, accustomed to laboratory work, attempted to preserve the scene using photographic slides. Before they could make their escape, the trio were apprehended by security. They revealed their Greenpeace affiliation when questioned, and, much to their relief, were summarily released. While Bennett and Curry remained to handle local media requests, Saunders headed to the CBC studios in Toronto to provide his account. The event was the lead story on the CBC television news and second on CTV; Bennett suggests that it made only the fourth page of the *Globe and Mail* because Ontario Hydro had time to initiate a defensive media campaign before the next edition was released.[43] The incident was dismissed in the provincial legislature by Energy Minister James Taylor, who referred to the Greenpeace members as "a few environmental streakers ... interested in a little publicity."[44]

On 18 July 1977, the provincial government gave Ontario Hydro approval to proceed with construction of its third nuclear power station, to be located at Darlington. This was particularly galling for anti-nuclear activists because the government granted the project an exemption from the Environmental Assessment Act. Energy Minister Taylor insisted that plans for the nuclear station were too far advanced when the act was passed in 1975 and that any delays would prove costly, but the Opposition was incensed, noting that Ontario Hydro's request for an exemption had been filed a year earlier, providing plenty of time for a proper hearing.[45] Greenpeace joined the chorus, stating in a press release that the government was "ignoring the potential danger to the environment and to public safety."[46] It announced that it would utilize non-violent action to halt construction at Darlington.

Greenpeace made good on its threat on 1 October 1977. Together with approximately sixty members of Save the Environment from Atomic Pollution, a Bowmanville-based anti-nuclear group, twelve members of Greenpeace Toronto marched from the local zoo to the site of the Darlington Nuclear Generating Station, where construction had recently begun. The inclement weather caused most protesters to return home shortly after arrival at the construction site, but the Greenpeace members pitched two tents, intent on forcing a confrontation. As an unidentified member informed a CBC reporter, "We'll stay here and impede construction as long as we can."[47] After refusing Ontario Hydro and police officials' requests to leave, the twelve were arrested and forcibly removed from the site. Charged with trespassing, they were released after promising to appear in court the following month and to keep away from all Ontario Hydro properties.[48]

After languishing in the background, the Toronto chapter of Greenpeace had in a short time risen to national prominence. Dan McDermott recalls with pride that "within a matter of a very few months we were kind of conspicuous in the media."[49] This newfound recognition came with a price, however. Despite its ability to attract attention, Greenpeace Toronto lacked expertise in fundraising, which meant that the bill for the summer campaign was largely borne by its core members. Saunders estimates that he invested $5,000 of his own money in the various activities.[50]

For its part, Energy Probe responded to Darlington's exemption from the Environmental Assessment Act by holding a mock hearing on the front steps of Queen's Park on 2 November 1977. Coinciding with long-time nuclear critic and MPP Donald MacDonald's introduction of a resolution to revoke the exemption,[51] this event demonstrates the changing character of Energy Probe. Since its origins as the Energy and Resources Project, it had always been the most thoroughly academic and policy-oriented component of the Pollution Probe Foundation. Now, as Pollution Probe emphasized its role behind the scenes, Energy Probe aimed to grab headlines with publicity stunts. Not only did it place a growing emphasis on the nuclear industry but it also found itself moving increasingly beyond reasoned debate and appealing to emotions. Further evidence of this was the publication in 1978 of *Everything You Wanted to Know about Nuclear Power (but Were Afraid to Find Out!)*. Written by Jan Marmorek (who later reverted to her maiden name, Jan McQuay) and co-financed by a variety of sources, including the Canadian Coalition for Nuclear Responsibility, Maurice Strong, and the United Church of Canada, it featured deliberately sensational section headings such as "Safety? What Safety?" "Nuclear Encounters of the Worst Kind," and "Psst ... Wanna Buy Some Plutonium?" as well as true-life horror stories such as one about an American shipping clerk who handled a package of liquid plutonium, which led to a gruesome series of amputations before his death from cancer five years later.[52]

By 1978 the tide appeared to be turning against nuclear energy in Ontario. Ontario Hydro had been battered by increasing criticism of its expansion plans, and the provincial government had stopped defending the Crown corporation. In April, Hydro formally approached Reuben Baetz, the recently appointed minister of energy, for policy direction. His response, that the agency should abandon nuclear power in favour of hydro, marked a significant policy reversal. This was compounded in September 1978 by the release of the Porter Commission's interim report, which argued that electricity demand in Ontario would grow at 4 percent per year, not the 7 percent on which Ontario Hydro had based its expansion plans. The interim report recommended that Hydro diversify its power generation infrastructure.[53] Energy Probe criticized the report, however, arguing that although "we are encouraged that Ontario Hydro's plans are to be modified downward, we are not at all satisfied that The Commission has gone far enough either in the conservation and renewable energy estimates or in its criticism of nuclear technology."[54]

Criticism of Ontario Hydro's business plans was compounded by heightened critiques of the safety of nuclear power in the aftermath of the 28 March 1979 partial core meltdown of the Three Mile Island Nuclear Generating Station in Pennsylvania. A central event in the erosion of public confidence in the safety of nuclear energy, the timing was particularly eerie, given the release twelve days earlier of *The China Syndrome*, a Hollywood blockbuster that featured a series of safety cover-ups at a fictitious nuclear reactor.[55] Three Mile Island provided Energy Probe with a unique promotional opportunity. Lawrence Solomon explains:

I had just come back from Europe and spent time with Danish environmental groups. They had a successful campaign to stop a nuclear plant in Denmark and they had a newsletter ... they distributed outside theatres or coffee shops or places where they thought they would find supportive people, and then their newsletter would get people to make quarterly pledges to support the cause. I came back from Denmark and planned to have the same kind of fundraising campaign for Energy Probe, and I had actually produced a little newsletter modelled on the Danish example. The day that it was finished, or almost finished, Three Mile Island occurred early in the morning. So I quickly put a new headline on it and ... we produced a bunch of these newsletters on our Gestetner machine. That afternoon we were distributing those newsletters as people were leaving *The China Syndrome* and it said "It's not just a movie – this actually happened." People leaving the movie were surprised ... The movie seemed prescient, so it was very confusing to people coming out of the movie being told ... something similar to that had just occurred that day ... The film had a big impact on theatre-goers so they were concerned about nuclear power as they left the theatre, and then they had the opportunity to sign up [for more information].[56]

In short order, volunteers were distributing the newsletters at theatres throughout the city. The near-disaster at Three Mile Island, coupled with the work of anti-nuclear activists such as those at Energy Probe, resulted in a marked transformation in the public's attitude towards the technology. Public opinion polls conducted by Gallup indicate that in September 1976, 41 percent of Canadians supported increasing the amount of nuclear power generated in the country, 20 percent wanted to maintain the current amount, 14 percent wanted to stop the generation of nuclear power altogether, and 25 percent were undecided. By May 1979, only 23 percent wanted to increase nuclear power generation in the country, 34 percent wanted to maintain the current level, 29 percent wanted to stop the generation of nuclear power, and just 14 percent were undecided.[57]

Three Mile Island also inspired the organization of what would become the largest anti-nuclear protest in Canadian history. In the early morning of 1 June 1979, three protesters acting independently of Toronto's ENGOs scaled the seven-foot barbed wire perimeter fence at the Darlington construction site. Once inside the perimeter, they scaled a two-hundred-foot transmission tower. Ignoring security's requests to come down, the three climbers, stocked with food and water, announced that they would descend from the tower only if Premier Davis halted construction and held public safety hearings on the development.[58] Security opted to allow the protesters to "cool off" overnight, but the following day saw a massive escalation. An estimated crowd of a thousand, organized by Greenpeace and local community groups, marched the two-mile stretch from Bowmanville to Darlington, carrying signs with slogans such as "Better active today than radio-active tomorrow" and "Hell No We Won't Glow." Described by Toronto *Star* reporters Ross Howard and John Munch as "an anti-nuclear Woodstock, dominated by young adults in halter tops and cut-offs, blue jeans and T-shirts, carrying placards, throwing Frisbees, and feeding babies," protesters were greeted with speeches by actors Barry Morse and Donald Sutherland, a collection of scientists opposed to nuclear energy, and a supportive letter from Toronto mayor John Sewell. The six-hour demonstration was punctuated by ten Greenpeace parachutists, five of whom landed inside the construction site, followed moments later by fifty-eight more protesters scaling the perimeter fence. Eventually sixty-six of the protesters were arrested for trespassing. As event organizer John Bennett of Greenpeace Toronto explained to the press, "It was a peaceful but unprecedented statement. A lot of people will read or learn that a lot of people already fear nuclear power."[59] This anti-nuclear action even received support from an unexpected source. William Peden, vice chairman of Toronto Hydro-Electric and a former staff member at Energy Probe, stated that the protesters merely wanted information and had "exhausted legal routes in trying to present their case and get answers to their questions." Noting that they were looking out for the public's interest, Peden added that "their views should be heard and I don't understand why the information is being withheld."[60]

In February 1980, the Porter Commission released its final report. Calling for a paradigm shift, its major conclusion was that Ontario Hydro ought to abandon its goal of increasing electrical output and the concomitant obsession with increasing the number of nuclear generating stations in the province, and instead focus on demand management – that is, focus on conservation and efficiency, and smaller-scale, low-impact energy generation projects. These recommendations echoed those long advocated by Energy Probe. Ontario Hydro initially chose to ignore the commission's recommendations, justifying its expansion by selling excess capacity. However, after the end of its contract with US-based General Public Utilities in June 1981, it ordered a slowdown of construction at Darlington and shelved the long-standing plans for additional nuclear facilities.[61]

FIGURE 14   In collaboration with local community groups, Greenpeace Toronto organized a major protest against the construction of the Darlington Nuclear Generating Station. The event, which occurred in June 1979, involved a two-mile march from Bowmanville to Darlington and was punctuated by a team of Greenpeace parachutists landing inside the closed construction zone. Additional protesters scaled the fence surrounding the site, leading to the arrest of sixty-six protesters for trespassing. Courtesy of Kai Millyard/Greenpeace.

Energy Probe Leaves the Pollution Probe Foundation

The Pollution Probe Foundation's economic troubles continued as the 1970s drew to a close. Between the 1977 and 1979 fiscal years, the foundation raised $307,326, or an average of just $102,442 per year. Monthly wages dropped from $750 in 1976 to $600 in 1980, which was roughly the same salary staff had been paid ten years earlier. To put this in perspective, in 1980, Statistics Canada's low-income cut-off (its equivalent of the poverty line) for a single adult living in Toronto was $7,008 after taxes. Even this uncompetitive wage was not guaranteed, however, and the foundation found it more and more difficult to meet its payroll.[62]

When Monte Hummel's tenure as executive director ended in May 1976, the organization had reverted to a flat, non-hierarchical structure. Although Pollution Probe and Energy Probe had a degree of operational autonomy under the Pollution Probe Foundation umbrella, members of both projects were expected to attend weekly meetings together. As with Pollution Probe's earlier experiment with collective operations, meetings were often sidetracked and became bogged down, but now the situation was exacerbated by financial difficulties and the result was an extremely trying workplace environment. As Pollution Probe employee David Coon recalls:

> The atmosphere at the organization at the time was horrendous ... Usually [at the weekly meetings] you'd have all the teams gathered there and it was not unusual for people to leave the room crying. It was just a very unpleasant working situation, kind of like a marriage that had gone very wrong. I remember one staff person bringing in knitting to try and stay calm – she knitted away through those meetings. Lots of passion and anger would break out regularly. It was not a good time.[63]

Energy Probe was faring considerably better than its peers at Pollution Probe and was debt-free.[64] This led Lawrence Solomon to advocate breaking off from the Pollution Probe Foundation. As he explains now, "the funding was terrible at the time. Salaries weren't being met. No one was really happy with the status quo." Solomon also cited differing approaches to the nuclear issue as a reason the groups should split:

> Pollution Probe received almost all of its funding from government and industry, and virtually nothing from the general public. Energy Probe received no money from government or industry, primarily because of Energy Probe's anti-nuclear position, which was very unpopular at the time. At the time, public opinion was loudly in favour of nuclear power ... and Energy Probe was an embarrassment to Pollution Probe. In fact, it [the Pollution Probe Foundation] didn't even let Energy Probe call itself "anti-nuclear" – it had to be "non-nuclear" ... There wasn't much reason to stay together, really.[65]

Other former staffers reject Solomon's assertion that nuclear energy was a wedge issue within the foundation; likewise, data collected by Gallup demonstrate that in May 1979, 63 percent of Canadians opposed the expansion of the country's nuclear generating capacity.[66] There is agreement, however, that Energy Probe's approach to fundraising did present problems. As David Coon explains: "Much of the friction was over fundraising time. The way we functioned at the time was we had three fundraisers on staff ... and I think there was a sense that, on the part of some people in Energy Probe, that they were devoting far more time to fundraising for Pollution Probe projects and activities than for Energy Probe. It appeared to me at the time that that was really the root of the internal conflict."[67] According to Jan McQuay and Lawrence Solomon, Energy Probe staff were particularly frustrated with the arrangement because it prevented them from pursuing a direct mail fundraising campaign, an innovative practice at the time, because it was considered unproven and too expensive.[68]

To assuage tensions, each project was granted greater autonomy. Pollution Probe staff and the *Probe Post* moved into the hitherto unopened Ecology House, while Energy Probe remained at the office at 43 Queen's Park Crescent. Besides more spacious quarters, Energy Probe also received permission to pursue its desired fundraising efforts.[69] As Marilyn Aarons of Energy Probe wrote to the Pollution Probe Foundation Board of Directors on 12 August 1980: "We are most pleased with the improvement of both morale and working conditions which has come about since we acquired some autonomy and cannot imagine it being in the foundation's interest to consider going back to the old arrangement."[70]

It soon became apparent, however, that Energy Probe could not be placated. On 24 September 1980, a majority of both staffs as well as the Board of Directors met to discuss a permanent solution to the Pollution Probe/Energy Probe structural problem. Two options were presented. Option A was the status quo: both groups would remain within the Pollution Probe Foundation but have fundraising and budgetary autonomy. Option B would see Energy Probe break away from the Pollution Probe Foundation and either join another established foundation or create its own. It was revealed during the meeting that Energy Probe had met with legal counsel earlier in the day and had begun the process of incorporating as the Energy Probe Foundation. It had also begun to assemble its own board and figure out how to divide assets. As board member Janet Wright recorded in the minutes:

When these conclusions were announced, there was considerable dismay on the part of some board members. It was felt that the board had not been sufficiently consulted, and that it was now being presented with a *fait accompli* and being asked to give rubber-stamp approval to a decision that had already been taken. On the other hand, a number of staff members asserted that if some action had not been taken immediately, the work of both PP and EP would have ground to a halt. The working relations between the two groups had deteriorated to such an extent that the work of all staff members was adversely affected.[71]

After a lengthy discussion, which Wright notes was "remarkably similar to a marriage counselling or divorce court session,"[72] a vote was held on Energy Probe's separation from the Pollution Probe Foundation. Nine staffers of both projects voted in favour and four abstained, while six board members voted in favour and one abstained. No one present voted against the proposal. The rest of the discussion centred on the status of the *Probe Post* – which remained a publication of the Pollution Probe Foundation – and the division of assets. The newly independent Energy Probe filed the paperwork for incorporation of the Energy Probe Research Foundation, which received charitable status in June 1981. As Solomon notes, there was discussion about changing its name to something more distinguishable from Pollution Probe, but the name Energy Probe was retained because it was "a brand people recognized" and "we also feared having to re-introduce ourselves with a brand new name."[73]

Morale rebounded in both organizations following the separation, which led to other significant changes. Pollution Probe's relocation to Ecology House meant the end of its affiliation with the University of Toronto, which dated back to its founding in 1969. More important, the split revealed Pollution Probe's underlying fiscal problems, which David Coon recalls came to a head at a board meeting shortly before the 17 October 1980 official opening of Ecology House:

> The board was looking at the books and they said, "Well, there's $40,000 in 'receivables' here. Where's the money coming from?" And I remember saying, "Well, we've got an agreement with the federal government under the Conservation Renewable Energy Demonstration Program to provide funds to finish off all the educational displays and so on at Ecology House." And they said "Where's the contract?" I said, "Well, we haven't gotten one yet." Then they said, "Okay, fine. The staff can leave right now, we're going to have a discussion."[74]

The financial naïveté of budgeting $40,000 – or roughly one-third of Pollution Probe's total 1980 revenue – without first securing a contract led the board to decide that the ENGO required the regular oversight of an executive director. This idea was wildly unpopular among the staff but, faced with the alternative of having their affairs micromanaged by the board, they grudgingly relented. The search for an executive director was launched in December 1980 and ended with the hiring in 1982 of Colin Isaacs, a former university administrator fresh from a stint as the New Democratic Party's environment critic at Queen's Park. Initial plans were for Isaacs to be a part-time executive director while holding another job at an education-oriented consulting firm, but Pollution Probe's deep-rooted financial problems soon led him to work at the ENGO full-time.[75]

Meanwhile, staff at the Energy Probe Research Foundation were enjoying a newfound sense of financial security that came with their independence. "One thing about Larry [Solomon] – he was very good at getting funds," recalls David Brooks. "He really made life a lot better for us because we were operating on pretty marginal salaries at the time."[76] In part this funding came from the implementation of the long-desired direct mail fundraising plan. Solomon, who claims that his was the first organization in Canada to adopt this fundraising method, explains that it was made possible by a member of Energy Probe's Board of Directors who headed Noma Industries' computer division and granted access to the machines during evenings and weekends.[77] Solomon also became adept at securing funds from controversial sources, such as the heavily polluting oil industry. In accepting such funds, Energy Probe crossed an important line. It is true that much of Pollution Probe's success could be attributed to its ability to secure support from industry; however, it also refused to grant immunity from its critiques in exchange for funding. Energy Probe, on the other hand, became a vociferous supporter of the oil industry, launching a campaign in 1983 "to educate Canadians to the social, environmental and economic benefits of less regulation in the petroleum field."[78] Boasting an endorsement from the Canadian Petroleum Association, it added hydro to the list of energy forms it opposed, leaving little doubt in the minds of its critics that the organization had compromised its credibility.[79] Solomon's ability to secure funding increased his influence within the organization, which in turn led Brooks to resign in frustration in 1982, after Solomon insisted that an employee working in Energy Probe's Ottawa office be fired.[80]

When Energy Probe left the Pollution Probe Foundation, its core staff consisted of Marilyn Aarons, David Brooks, Chris Conway, Jan Marmorek,

FIGURE 15 Staff at Energy Probe, following their split from the Pollution Probe Foundation. Standing (left to right) are Pat Adams, Jan Marmorek, Jack Gibbons, and Marilyn Aarons. Seated (left to right) are Norm Rubin, David Brooks, and Lawrence Solomon. *Source:* Personal collection, Marilyn Aarons.

Norm Rubin, and Lawrence Solomon. By 1983 just Aarons, Rubin, and Solomon remained. Marmorek's departure was hardly acrimonious – she separated from her husband and required more income than could be found in the low-paying world of Canadian ENGOs – but Conway had grown weary of Energy Probe's new direction. He disagreed with the "fear mongering" tone it had adopted on nuclear issues, as well as the increasing focus on free market solutions.[81] These staffers were replaced by individuals who accepted Solomon and Rubin's belief in free markets and deregulation as the solution to society's problems. Energy Probe, whose staff had long espoused a mélange of ideological positions, evolved into a libertarian stronghold, as was clearly illustrated in Solomon's 1984 publication, *Breaking Up Ontario Hydro's Monopoly*, which argued the case for privatizing the province's publicly owned utilities provider. Energy Probe's overtly ideological positioning within the environmental movement troubled many of its peers, who favoured more government intervention and feared that reducing regulations would give free rein to polluters. Thus, although Energy Probe's suspicion of government planners was in line with the neoliberal agenda that swept Margaret Thatcher and Ronald Reagan to office, it also caused the organization to be viewed with suspicion by the broader environmental community.

Recycling

Pollution Probe had long led the push to popularize recycling in Toronto and throughout the province of Ontario. To this end, it had organized pilot projects, served on government committees, and held public demonstrations. It had also created the concept of the 3Rs waste hierarchy, which became synonymous with the practice of recycling. In August 1978, it released "Probe's Last Word On 'Recycling,'" a nine-page overview, arguing that "it's time we spent our effort on the many other issues that need attention."[82] Retiring the issue was acceptable "now that action is underway," it said, referring to the opening that month of the mechanized Centre for Resource Recovery in Downsview, the first of six plants announced by the provincial government in 1974.[83] What was not mentioned was the fact that the decision was partially inspired by the ENGO's sparse coffers, which forced it to streamline its areas of interest. Still, the decision to abandon recycling was a risky one. On the one hand, the Centre for Resource Recovery, which was designed, in the words of Harold Crooks, "to shred, air-separate, cyclone, separate and load 900 tons of refuse for transportation to waiting markets – all in two eight-hour shifts,"[84] could turn out to be a great success. On the other hand, Pollution Probe freely admitted that the technology had not yet been perfected. Should the experiment fail, Pollution Probe, which was renowned for its expertise in recycling, would have already capitulated on the issue. As it turned out, the Centre for Resource Recovery was an unmitigated failure, and the plant, as well as plans for additional recycling facilities in the province, was abandoned.[85]

Even as Pollution Probe was declaring victory and moving on, the push for recycling was picked up by the Is Five Foundation (IFF), which was launched in October 1974. Its founder, Jack McGinnis, had a degree in communications and had pursued a career in photojournalism. While he was concerned with waste, inefficiency, and disrespect for people and the environment, McGinnis did not actually set out to develop an environmental organization, but, rather, a workers' cooperative.[86]

The Is Five Foundation's unusual name was deliberate, as McGinnis felt it would provide a natural opportunity to explain the organization's purpose. It was derived from Buckminster Fuller's concept of synergy and E.E. Cummings's book of poetry *is 5*. According to McGinnis, "the idea was to find a way for people to work together so that it was exciting and inspiring, and so ultimately the whole would be greater than the sum of the parts, and what we did together would be more than if we worked on

FIGURE 16   The Is Five Foundation operated curbside recycling programs in Toronto during the late 1970s. Courtesy of Is Five Foundation.

our own." In essence, the aim was to empower people through cooperation: "We wanted to tell people there was a problem, but the solution was them in their own home and their own lifestyle. So it was very much people working together within the group, and trying to find practical ways to ask people in their own home and eventually in their workplace to do things differently."[87]

The IFF was established as a nonprofit, registered charity and began operation as a collective, with its seven initial members all participating in the decision-making process. Its first effort, a roadside multi-material pickup that operated weekly in the east-end Beaches district – known as Project One Recycling – was launched in January 1975. Members travelled from door to door publicizing the program, while McGinnis drove the organization's lone vehicle, a pickup truck. According to the IFF, Project One Recycling was a matter of practical research. The numbers were not particularly impressive – an estimated four thousand participating residents by 1977 – but McGinnis was generally pleased with the results. As he later noted: "We didn't have professional equipment. We didn't have blue boxes. Everybody had to use cardboard boxes or whatever. So there were definitely limits. What went well was the community involvement and the fact that people would listen to reason. People were proving what we believed in: people were naturally good, you just needed to give them the tools." The IFF would later discover that this was the first roadside multi-material pickup to operate in Canada.[88]

The IFF's approach to recycling represented a philosophical break from Pollution Probe. Whereas the latter had trumpeted mechanical separation plants as the only realistic way to address recycling in the city, Is Five believed that separation at source was essential, as it would force participants to consider their consumer habits. Derek Stephenson, the organization's research coordinator, described the necessity of active participation to *Globe and Mail* reporter John Marshall: "Individuals just can't see how they can clean up the Great Lakes, save the seals, stop rip-offs. But they can peel labels off cans. It's a start towards an acceptance of the environmental ethics of a conserver society."[89]

McGinnis's astute business sense enabled the IFF to expand dramatically in its second year. Seeking support from the Local Initiatives Program, he recognized that there would be major competition for funding, which was capped at $100,000:

> We knew we were up against a lot of competition after our first year because other people had heard about the program and even though we'd done fairly well and they seemed to like what we'd done in year one we knew we'd have to be clever. And we wanted to get bigger and figured out they gave out the money riding by riding. So there was competition between a federal riding, but often there was a bit of money left over once they got done deciding who was going to get the priority. So we figured out how to come up with the smallest grant we could apply for – the least amount of people for the shortest amount of time. I did twenty-one applications, photocopied exactly the same with every federal riding in Toronto, except the one in the Beaches where we had our original grant. So with the Beaches we got another round of seven people as the head office, and out of the twenty-one [applications] we submitted they approved eleven of them, without knowing it. When they had their first get-togethers for the project officers to meet their new grantees, it was only then that they figured out how much money they'd given *[laughs]*, which was well over $100,000.

McGinnis's canny manoeuvring, which netted just over $129,000, led to a revamped application process the following year, with LIP applicants being required to identify whether they were applying for funding in any other federal riding.[90]

The LIP funds enabled Is Five to undertake a variety of projects, including a study of "traffic calming patterns" in Christie Pits. However, although the IFF momentarily found itself to be Canada's largest environmental organization, with twenty-nine full-time employees, its finances remained unstable. When the LIP grants dried up in the summer of 1976, staff were forced to fund their work with personal savings and income drawn from alternative employment while they awaited the results of funding applications. The ongoing quest for financial stability led the IFF members to explore a plethora of money-making ventures, including woodworking, graphic design, and the operation of a printing press.[91] The solution to the IFF's financial woes soon appeared, however. As Stephenson explains: "We were starting to get lots of consultants, people in really nice suits, coming by our operation to learn how we were doing things. We would tell everybody everything."[92] The IFF subsequently decided to form a consulting firm specializing in waste management and recovery systems, Resource Integration Systems (RIS), which was launched in March 1977. With Stephenson as president, RIS funded the IFF's activities by charging consultants' rates for its expertise.

On 8 December 1977, the IFF submitted a proposal to the East York Works Committee to operate a weekly newspaper pickup throughout the borough, with the IFF assuming all costs. Approved by the committee within four days, the plan received the go-ahead from East York Council on 19 December.[93] Operating under the auspices of the East York Conservation Centre, pickup began in February 1978, with two trucks. Six months later, the program achieved 33 percent participation, averaging twenty-five to thirty tons of newspaper per week. By June 1979, participation had increased to 45 percent and collections to thirty-five tons per week. Now "Canada's largest non-municipal source separate waste reclamation program," the East York Conservation Centre had moved from collecting only newspaper to cardboard, glass, and metals.[94] A November 1979 report explained that the "project was initiated to provide a demonstration of the viability of local at-source recovery programs."[95] Documenting their extensive planning in a series of reports, the IFF also used the opportunity to study the functionality of various technologies and approaches to recycling. As had been identified early on, a "major barrier to the successful implementation of at-source recovery on a broad scale was ... a lack of suitable collection equipment designed for multi-material curbside collection of recyclable materials."[96] The IFF had started with a pickup truck in the Beaches in 1975; by the time the East York pickup began, it had purchased a GMC MagnaVan with a 2.5-ton carrying capacity and was renting a similarly equipped vehicle. Funding from Environment Canada's Development and Demonstration of Resource and Energy Conservation Technology Program enabled the IFF to collaborate with Toronto-based DEL Equipment in creating a vehicle specially designed for recycling programs. The prototype reduced the physical labour involved in collection, enabled a two-person crew to collect multiple waste streams, was capable of automatic unloading, and was competitively priced with existing collection vehicles.

In 1978 Jack McGinnis secured a grant to spend three months meeting with recycling advocates and practitioners throughout Ontario in order to determine the need for a provincewide recycling organization. He also visited the West Coast to examine the model of the British Columbia Recycling Council, which was formed in 1973. This resulted in the launch of the Recycling Council of Ontario (RCO) in June 1978, operating out of the IFF offices at 477 Dupont Street. In October 1978, the council received $15,000 from the Ministry of the Environment and Eric Hellman, an RIS staffer, was hired as its executive director. The RCO had a twofold agenda: to serve as a network for the province's nonprofit recycling groups, and to develop cooperative marketing for its members. It had an early brush with success when the Ontario Paper Company decided to build a de-inking plant in Thorold. The RCO had offered to provide 64 percent of the plant's needs within three years, but an unstable market and pressure from the province's traditional paper companies, which now viewed the organization as a threat, led the RCO to abandon its cooperative marketing directive. Despite this, the RCO flourished as an information provider. In March 1981, it established the Ontario Recycling Information Service, which created a toll-free telephone line to answer the public's queries about recycling and available programs. By 1990 the service was fielding 20,500 questions annually.[97]

The IFF made further progress with its consulting arm, RIS. In July 1977, it received a subcontract to design and implement a multi-material recycling program for Canadian Forces Base Borden. The project was conceptualized by Rick Findlay, senior project engineer at Environment Canada's Environmental Protection Service, and was inspired by a visit to the still-unfinished Centre for Resource Recovery in Downsview. Noting that separation at source would be much more efficient than the unproven mechanical separation system in which the province had invested $20 million, Findlay chose the military base because of its proximity to markets for recovered materials, the detailed knowledge of its past waste generation and management practices, and the Department of National

Defence's amenability to the project.[98] As Derek Stephenson recalls: "We were essentially given this place to experiment with recycling. Had a good budget, but we were subcontractors to consultants who were theoretical, MBA-types, while we were operational types. And from that experience we both got to play around with other people's money and perfected a lot of techniques."[99] The project involved collection of corrugated boxes from base shopping centres, glass and bottles from its drinking establishments, paper and newsprint from its offices, and cans, newspapers, and glass from its residences. When it ended in March 1979, the project was considered a success, with 45.9 percent participation in the curbside collection of newspapers and 21.4 percent for glass. It was determined that this program, if continued, could provide over $15,000 in net profit annually.[100]

Conclusion

The late 1970s was a time of great turmoil for the Pollution Probe Foundation. Beset by financial difficulties, it saw its long-time anchor, Pollution Probe, struggle to assert its relevance at a time when interest in environmental issues was at its nadir. These difficulties inspired the more successful Energy Probe to depart the Pollution Probe Foundation in 1980.

The same period also witnessed a transformation in the Toronto environmental community. Once dominated by Pollution Probe and its institutional offspring, it was now populated by unrelated groups that thrived in specific niches. Greenpeace Toronto established itself as the city's pre-eminent action-oriented ENGO, in sharp contrast to Pollution Probe's focus on policy work. The Is Five Foundation, meanwhile, applied a highly effective hands-on approach to the recycling issue. Combined with the emergence of the independent, free market-oriented Energy Probe, these developments spelled the end of Pollution Probe's leadership of the local environmental community. As the 1970s gave way to the 1980s, the organization that had shaped that community would be forced to re-evaluate itself in the context of the broader movement, or risk going the way of GASP.

6
Beyond the First Wave

Pollution Probe returned to the forefront of the public consciousness in the summer of 1989. The Loblaws supermarket chain, eager to capitalize on the new trend of green merchandising, launched its own line of "environmentally friendly" products. In order to legitimize its Green Line in the eyes of consumers, Loblaws struck an agreement with the two-decade-old environmentalist organization. To Pollution Probe executive director Colin Isaacs, this provided an ideal opportunity. Free to pick and choose which items it would endorse, the organization would help educate consumers while earning a 1 percent royalty on sales of these products. Furthermore, Pollution Probe's endorsements would be featured in a national advertising campaign, providing a boost to its public profile.

Involvement in Loblaws's Green Line promised to be the capstone of Pollution Probe's decade of revitalization. The organization emerged from the late 1970s in disarray. Leaderless and without a clear direction, it also found itself perpetually short of funding. Under Isaacs's entrepreneurial leadership, Pollution Probe was able to address both problems.

This organizational turnaround coincided with a realignment of the Canadian environmental movement. Whereas localized groups had long been the norm, the second wave of the environmental movement, beginning in the 1980s, saw the rise of pan-Canadian environmental organizations. These groups proved to be particularly adept at addressing the emerging transnational environmental concerns of the era. It would be wrong, however, to think that the end of the first wave coincided with the demise of the earlier groups. This chapter examines the ongoing activities of Pollution Probe, Energy Probe, and the Is Five Foundation spinoff, Resource Integration Systems (RIS), amid the changing landscape of the Canadian environmental movement.

Pollution Probe's Revival

Pollution Probe's return to form began in the early 1980s, in large part due to a newfound focus on hazardous waste and public health. These interconnected issues emerged in the aftermath of the Love Canal disaster, in which residents of Niagara Falls, New York, discovered that they were living on sites contaminated by waste from the Hooker Chemical Company. This had a direct bearing on Canadian interests, as it was soon discovered that Hooker had four hazardous waste sites that were leaking into the Niagara River and ultimately Lake Ontario, the source of drinking water for an estimated four million Canadians. Despite the transnational nature of this environmental issue, historians have ignored the contributions made by Canadian ENGOs to its resolution. The idea that these hazardous waste cases were strictly American affairs highlights their impact as a human interest story. Seen in this light, it was the story of middle-class Americans whose life savings were jeopardized when they unknowingly purchased homes on contaminated land. The prospect of buying a home, a key component of the American dream, only to discover that it might have lethal consequences, evoked strong emotions from the public. This overshadowed the more abstract aspect of the story, namely, that its environmental consequences crossed international boundaries.[1]

Pollution Probe waded into the issue in May 1981, when it was revealed that an out-of-court settlement had been reached between the United States and New York governments and Hooker concerning the latter's waste dump at Hyde Park Boulevard. According to the terms of the deal, Hooker would spend $15.5 million capping the site, collecting any chemicals that escaped in drainage pipes surrounding the site, and cleaning the nearby Bloody Run Creek and Niagara Gorge. This treatment of the waste site, which contained eighty thousand tons of chemicals, including nine hundred kilograms of the highly toxic dioxin, was deemed wholly inadequate by Pollution Probe and Operation Clean – Niagara, a Niagara-on-the-Lake-based citizens' group. Through Toby Vigod of the Canadian Environmental Law Association (CELA), the two filed a request to intervene in a judicial overview of the proposed settlement.[2] Vigod argued that her clients were requesting *amicus curiae* (adviser to the court) status because they "feel that the settlement agreement must contain conditions stringent enough to ensure that international waters are not contaminated."[3] The request was accepted and Pollution Probe and Operation Clean – Niagara were given an opportunity to review the terms and to submit their comments to the court. The submission, signed by Vigod on behalf of her clients and by Barbara Morrison, an American attorney representing the New York-based Ecumenical Task Force, argued that the current agreement would leave the site contaminated, which would have profound implications for millions of Canadians and Americans whose drinking water source would be rendered "extremely toxic." It argued, therefore, that the best solution would be to excavate and destroy the waste, with re-entombing it in a secure vault a distant second choice. This carefully prepared brief led Judge John Curtin to hold hearings featuring expert witnesses, including Grant Anderson, a Canadian hydrologist, and Douglas Hallett, of the Canadian Wildlife Service, both of whom supported the conclusions found in the Vigod-Morrison brief. Curtin finally approved the settlement in April 1982 without acceding to the environmentalists' calls for a wholesale change in plan. Nonetheless, historian Elizabeth Blum argues that the settlement did include some important provisions, including remedial work on the site and a requirement that the Occidental Chemical Corporation, Hooker Chemical's parent company, identify the scope of the contamination within the community.[4]

Pollution Probe received important support from government officials during its intervention in the Hyde Park Boulevard settlement. Although John Roberts, the federal minister of the environment, opted to avoid direct involvement, he wrote a letter of support for the *amicus curiae* bid. Keith Norton, Ontario's minister of the environment, met with the group to offer advice.[5] When attention then turned to the so-called S-area waste dump, Pollution Probe requested that it and Operation Clean – Niagara receive intervener status, which would give them the ability to introduce evidence and question witnesses on a level playing field with Occidental Chemical and the various levels of government involved. This would be a clear step up from *amicus curiae,* which only grants permission to supply information to the proceedings if so requested by an intervening party. In October 1982, however, Norton announced the provincial government's intention to intervene, stating that this was rooted in its desire "to have the maximum influence in any decision made." It appeared to be a reaction to the public interest that had been spurred by the Ontario-based ENGOs' involvement, and Pollution Probe, which had already secured

$35,000 in research funds for the case from Environment Canada, publicly requested that the Ontario government back out in order to avoid contradictory evidence.[6] Nonetheless, when the time came for the court to review the proposed settlement, the government of Ontario was granted intervener status while Pollution Probe and Operation Clean – Niagara were relegated to *amicus curiae* status. A clearly nonplussed Colin Isaacs told the media that he did not trust the province "to protect the health and safety of the people of the Niagara frontier or to protect the waters of Lake Ontario from a landfill that spews toxic chemicals into the Niagara River."[7] When the court hearings began in May 1984, Pollution Probe was highly critical of the government's handling of the case, which featured just two witnesses, both of whom admitted under cross-examination that they had not read the proposed settlement, and a Washington-based lawyer, Philip Sunderland, whose ill-prepared case was described as both "contradictory" and "silly" by the presiding judge.[8] Isaacs opined that the Ontario Ministry of the Environment "were made a laughingstock and it is terribly damaging to the ministry's credibility." Pollution Probe's critique worked its way back to the provincial legislature, where Norton's replacement as environment minister, Andrew Brandt, was forced to defend the government's performance.[9]

While Pollution Probe played a vital role in ensuring that Canadians' voices were heard in the Hooker Chemical settlements, its greatest impact came as a result of its involvement with SCA Chemical Waste Services, a company that assisted industry in the disposal of chemical wastes. In January 1980, SCA was granted permission by the New York Department of Environmental Conservation to build a five-and-a-half-mile pipeline to dump treated chemical waste into the Niagara River. The Ontario government indicated that it was not concerned with the decision, but members of Pollution Probe, Operation Clean – Niagara, and the Niagara Falls, New York-based Operation Clean announced their objections and their intention to appeal. In April 1982, SCA, Pollution Probe, Operation Clean – Niagara, and Operation Clean reached an agreement that would see an increased level of monitoring, including a study that would chart the path of the wastes as they became diluted, in exchange for an end to the environmentalists' opposition. Anne Wordsworth of Pollution Probe noted that her organization continued to oppose the dumping of wastes but supported the agreement since it imposed a rigorous set of controls while at the same time reserving the rights of Pollution Probe, Operation Clean – Niagara, and Operation Clean to reopen hearings if they were dissatisfied with SCA's performance.[10]

This agreement had profound implications. At the time, SCA's 5.6-acre landfill was approaching its capacity, so the company proposed the creation of a new 25-acre landfill site that would be capable of holding more than a million tons of industrial waste. Under the terms of the April agreement, this proposal had to be cleared by the Citizens' Review Board, which had representatives from SCA, the three environmental groups, and the New York state communities of Porter and Lewiston. As lawyer Barbara Morrison explained in a letter to Judge Francis Serbent:

> The contaminants from the landfill will discharge into Six Mile Swale/Four Mile Creek, and potentially into Twelve Mile Creek – these three streams flow into Lake Ontario; hydrogeologic considerations, monitoring plans and air emission calculations are inadequate; and there are potential environmental impacts to the Niagara River and Lake Ontario which may result from construction and operation of the proposed landfill and may continue as a major problem after closure of the landfill facility.[11]

Serbent ruled that SCA could proceed without hearings, but his decision was overturned in an appeal to the New York Department of Environmental Conservation, whose commissioner found that the issues of monitoring, leachate compatibility, and air emissions required further examination.[12] Prior to the launch of public hearings, representatives of Pollution Probe and SCA met to discuss possible solutions; these discussions soon expanded to include the Department of Environmental Conservation. The resulting agreement, first announced on 6 February 1984, would phase out landfilling of the most hazardous wastes. As Pollution Probe later boasted: "It is believed that it ... marks the first time in North America that citizen group opposition to a landfill has led, through multi-party discussions, to implementation of an environmentally preferred solution not only in the location of initial concern but also throughout a legislative jurisdiction."[13] In turn, Pollution Probe would pressure the Ontario government to follow New York's lead, arguing that similar legislation in the province would divert 13 million litres of hazardous waste from public dumps, 18 million litres from private landfills, and another 13.5 million litres that were poured into municipal sewers.[14]

Even as it worked on the toxic waste issue, Pollution Probe turned its attention to the safety of Toronto's water supply. This first made headlines in November 1981 when Anne Wordsworth presented a report to the Toronto Board of Health that questioned the long-term effect of the low-level toxins detected but permitted under Environment Canada guidelines. This led an employee of the Toronto Board of Health to warn pregnant women to avoid drinking tap water – a comment that was immediately rebuffed by Alexander Macpherson, the city's medical officer of health, and Metro Chairman Paul Godfrey, who accused Pollution Probe of fearmongering.[15] A follow-up report released in February 1982, in which Pollution Probe urged further research into the long-term effects of the low levels of benzene detected in the Metro water supply, led Godfrey to not only dismiss the findings but also encourage the public to stop supporting the ENGO in response to this "sure sign of irresponsibility."[16] Pollution Probe again critiqued the city's water system in March 1983, with the release of *Drinking Water: Make It Safe*. This research paper alleged that fifty-three contaminants were found in the water supply – including sixteen known carcinogens – and claimed that between 72 and 156 Metro residents would develop cancer in their lifetime as a result of the polluted water supply. This could be rectified, it argued, by the addition of carbon-based water purification systems, at an added cost of $8 to $16 per Metro resident per year.[17] The peculiar exactitude of Pollution Probe's claims brought the study under intense scrutiny. Although Pollution Probe was quick to defend itself, noting that its work was based on published data, George Becking, chief of environmental toxicology at the federal Department of Health, alleged that the cancer estimates were based on disputed research. Becking delivered a mixed message on the matter, however, noting that "there's no reason to consider that there is a long-term excessive risk from drinking Toronto water" while simultaneously refusing to call it "safe," given the lack of long-term research on the subject. Godfrey responded by sipping a glass of water for photographers, adding that "I would ... let my kids drink it by the barrelful."[18] Frustrated that residents were switching to bottled water while the municipal supply was "getting a black eye," Metro Works Commissioner Frank Horgan announced the following month that his department would begin subjecting bottled water to the same chemical analyses applied to the city's water. While Metro held fast to its position that the water supply was safe, the provincial Ministry of the Environment announced in June 1983 the creation of an internal panel of experts on water toxins, with a focus on dioxin, and a carbon filtration plant to test the technology in Niagara Falls.[19]

Pollution Probe's focus on toxic waste and the safety of the water supply was central to its rebound from the doldrums of the late 1970s.[20] These issues not only revived the organization's public profile but also served as the focus of new fundraising efforts. On arrival at Pollution Probe, executive director Colin Isaacs had concentrated his energies on securing additional funds from government, foundations, and corporations. He "quickly found that we were pretty much at the limit there of what we could raise."²¹ The inability to coax additional money from these sources is understandable, given the severe economic recession in Canada in the early 1980s. However, the fact that Pollution Probe, over a decade after its founding, was still relying on the same three sources of funding was a clear indication of the leadership vacuum in recent years. Isaacs therefore turned his attention towards the general public, a source of revenue long untapped by the organization. This brought Pollution Probe to the world of professional fundraising, with an emphasis on direct mail campaigns and door-to-door canvassing. Although not an ideal solution because of its high cost, it did result in a significant increase in the Pollution Probe Foundation's revenues, from a low of $86,022 in 1978 to an average of about $300,000 between 1981 and 1984, which in turn meant no more missed paydays.²²

The addition of paid fundraisers had other implications for Pollution Probe. Decisions concerning operations were still largely made as a collective during weekly meetings. "That was okay when the staff was eight or nine or ten," explains Isaacs, but the influx of fundraisers pushed the number of staff towards fifty. "It became totally unmanageable so I moved towards a system where individual teams would make the decisions for their teams and I would act as a mediator when there was conflict between teams."²³ The board of directors approved this change but it was controversial among the staff, who felt that it gave Isaacs too much power over the organization.²⁴

That Pollution Probe survived to celebrate its fifteenth anniversary was a significant accomplishment. Between 1969 and 1971, ENGOs had emerged in cities across Canada, but very few of them survived to the mid-1980s. Pollution Probe's longevity was in many ways connected to its low overhead, thanks to rent-free office space at the University of Toronto through 1980 and dollar-a-year rent thereafter at Ecology House. And despite a reduction in available money for much of this period, it was effective at securing funds to continue operations. The same cannot be said of its namesake affiliates. Whereas fifty existed in the province of Ontario in 1971, just one, in Ottawa, continued to operate into the 1980s, at which time it was "bought out" by Pollution Probe in Toronto.²⁵

The Green Line Fiasco

In 1988, just as the Toronto real estate market began to take off, Colin Isaacs convinced the federal government to sell Pollution Probe's long-time

FIGURE 17   Pollution Probe's new home, Ecology House, opened in 1980. Designed as a showcase for conservation and alternative energy technologies, Pollution Probe rented the building from the federal government for a dollar a year until 1988, when the ENGO purchased it for $175,000 – a price well below its book value. Note the use of solar power, passive and active, throughout the building. *Source:* Personal collection, David Coon.

home, Ecology House, to the ENGO for $175,000 – a bargain for a building valued at over $600,000. While Isaacs saw this as a simple way to increase Pollution Probe's assets, other staffers chafed at the one condition: that a letter praising the federal government for the deal and its environmental record be forwarded to everyone on their seventeen-thousand-name mailing list. Isaacs noted that "if the minister hadn't asked, we wouldn't have suggested it," and also explained to the press that "I'm sure our members and supporters have the resources to interpret it."[26]

Internal strife, which had been growing, came to a head as a result of the organization's partnership with Loblaws's Green Line. While numerous items were endorsed, Isaacs landed in hot water for the measured approval he gave the company's "environmentally friendly" disposable diapers. In a television commercial, he took the stance that those truly concerned with the state of the environment would use cloth diapers; however, given the reluctance of many consumers to pursue this option, the fact that the diapers in question were biodegradable, used fewer trees, and were chlorine-free made them the preferred second option. Despite the qualified nature of the endorsement, two Pollution Probe staffers promptly resigned, citing discomfort with the decision to endorse disposable diapers in particular, and the concept of product endorsements in general; another three threatened to follow suit. Under pressure from the Board of Directors, Isaacs tendered his resignation, Pollution Probe withdrew from the Green Line program, and whispers of a mass revolt within the ENGO were stayed.[27] Loblaws continued the Green Line despite the unwillingness of other ENGOs to lend their names to it, proving that retailers were willing to market ecologically friendly products, but many in the target audience remained skeptical of their intentions.

The Green Line fiasco launched a public debate about the propriety of ENGO/corporate relations. In some it inspired an automatic, vehement denunciation. Others were more measured in their critiques. As Clifford Maynes wrote in a letter to the *Globe and Mail:*

> There is nothing wrong in the environmental movement "working with business" by publishing criteria for environmentally acceptable products, rating available products according to these criteria and advising business how to make improvements ... However, individual product endorsements are another matter. They imply that a particular product is the best choice or the only acceptable choice in the interests of the environment, when it may be neither.[28]

However, Isaacs reminded people that Pollution Probe had a history of collaborating with the business community, writing in the *Globe and Mail:* "For 20 years, Pollution Probe has sought access to the boardrooms of the nation, first to implement policy and second to raise money. The fact that we have solicited and accepted corporate donations seems to have taken people by surprise, even though it is published regularly in our annual report."[29] He might even have pointed out that there had been a previous collaboration with the grocery chain in the late 1970s that resulted in the sale of packaging-free products.³⁰ In a recent interview, Isaacs maintained his support for the partnership: "I believe that it was absolutely the right thing to do, that it helped to raise Pollution Probe's profile even further, and it helped to illustrate that environmental groups can work constructively with business." He blamed the fallout from this program for a chilling of relations between the two sectors that continues to linger.³¹

Two things about this event strike the historian as odd. First, it attracted more attention than any of Pollution Probe's environmental campaigns during the 1980s – a significant feat when one considers that the ENGO engaged in the toxic waste issue in New York state and questioned the safety of Toronto's municipal water supply. Furthermore, this marked the first time that Pollution Probe's funding sources came under public scrutiny. From the time of its inception, the ENGO had relied on funding from government, corporations, and foundations to pursue its various activities, but these sources, and their potential influence over Pollution Probe's actions, went unquestioned for twenty years. In large part, this was because the ENGO had never before involved itself in a self-serving endorsement of a product with such dubious environmental credentials. When Pollution Probe had come out in favour of certain items in the past, most notably when it encouraged the public to purchase detergents with low phosphate content in 1970, it did not receive royalties for its work. Pollution Probe's sponsors were always listed in its annual reports and mentioned in the appropriate press releases, but because the organization did not compromise its willingness to critique corporations and the government, its integrity was never questioned.

Expansion of the Energy Probe Research Foundation

The Energy Probe Research Foundation continued to carve out its eco-capitalist niche throughout the 1980s. In a move that saw it expand beyond its original focus on the Canadian energy sector, it became a fierce critic of Canadian foreign development policy, arguing that "foreign energy affairs [are] inseparable from domestic energy considerations."³² Energy Probe's first target was the use of Canadian International Development Agency funding to build hydro dams in Haiti. Energy Probe was quick to point out that the dams would cause the relocation of thousands of poor Haitian farmers, but its opposition was ideologically driven, as it viewed the use of Canadian taxpayers' dollars for foreign aid to be antithetical to free market principles. It argued that the infusion of foreign aid acts as a hindrance to Third World countries because it results in economically unviable megaprojects, which subsequently harm their economies and environments.[33] In 1986, finding that many of its supporters were confused by its interest in overseas matters, it created a separate Probe International project under the Energy Probe Research Foundation umbrella. This reorganization resulted in a minor boon for, as Solomon explains, "we found out everyone who was sending us a $25 cheque for Energy Probe would send us a $25 cheque for Energy Probe and Probe International. So our revenue pretty much doubled just by ... rebranding our international work."[34]

In 1988 the Energy Probe Research Foundation stood alone in the Canadian environmental community in supporting the Canada-United States Free Trade Agreement. In the wake of its passage, it launched Environment Probe, which focused on applying market mechanisms to benefit Canada's natural resource sector. More recently, it launched the Urban Renaissance Institute to study regulatory matters as they affect cities. As Solomon explained in an interview, "[we] always intended to become an all-purpose environmental organization."[35] Solomon's interest in the marketplace was not confined to policy work. In the late 1980s, he established a short-lived mutual fund that invested in ecologically friendly companies, and in 2004 he founded the Green Beanery, a café and coffee equipment store located in the Harbord Village neighbourhood, the profits of which fund the work of Probe International.

The Birth of the Blue Box

A key event in Ontario's history of waste disposal occurred in autumn 1977 when the Pollution Probe affiliate in Kitchener-Waterloo held a day-long event called Garbage Fest 77. Besides bringing together many of the province's foremost environmentalists, including the staff of the Is Five Foundation, it featured a speech from Nyle Ludolph, the director of special projects at Superior Sanitation. Ludolph had little interest in recycling before this event, but a day spent in the company of recycling advocates transformed him. "My conscience got a hold of me and I said 'I'm going to try this,'" remembers Ludolph. "I went home that day and dug up a hole in the backyard for compost, and I put boxes at the side door in the garage and I said to the family 'We're going to test this recycling thing.' Consequently, we ... only generated 102 lbs of garbage for the entire year."[36] This amazed Ludolph, who notes that the average family of three would normally generate a ton of garbage in that timeframe. Given the growing difficulty of acquiring landfill sites, he saw a way to help the company while at the same time earning public support. His boss, Ron Murray, president of Laidlaw Waste Systems, was also intrigued by the potential but was concerned about the business implications. Ludolph recalls: "He said, 'Look, if we do that we may as well park the garbage trucks.' And I said, 'No, no. For every garbage truck we take off we put on a recycling truck. What's the difference?' He kind of agreed with that concept. We weren't going to hurt our business any – it would complement our business."[37]

After the success of RIS at Camp Borden, Ludolph approached Eric Hellman about bringing recycling to Kitchener. According to Hellman, "He said to me 'Wouldn't it be amazing if we could do this citywide? If everybody would do this?' And I'm looking at this guy who was head of garbage collection for this company going 'Do I hear what I'm hearing? Does he actually want to do recycling?'" Hellman promptly wrote a proposal for a test program. He recalls:

> In the conversation about the proposal we had made to them he [Ron Murray] said something very frank. "We make our money off of garbage. We make a good living. But something in me says this can't last forever, that it doesn't make sense, business-wise or social-wise, to be paying somebody to keep picking up garbage. At some point this has to turn into something like recycling, where there's some good being made out of this material."[38]

Hellman's proposal to examine the efficiency of a variety of collection methods from a sampling of one thousand homes in Kitchener received $72,000 in funding from Laidlaw. RIS was given the opportunity to design the project, which would be carried out by Total Recycling, a new division of Laidlaw headed by Ludolph. The project was a great success. Originally scheduled for six months, beginning in September 1981, it continued uninterrupted until 1983, when the recycling program went citywide. Particularly encouraging results emerged from the one-quarter of homes that were given blue boxes in which to place their recyclables. This hardly surprised Ludolph, who had examined recycling programs in California while preparing for the test in Kitchener. In California each household used three bins, which were designed to separate the materials, but recyclers still had to sort materials from these bins. For the sake of simplicity, the Ontario project opted to utilize a single bin.[39] It was also at this stage that the now-iconic boxes acquired their blue colour. Derek Stephenson recalls that "with plastics the darker it is the less likely it [the colour] will break

Beyond the First Wave

FIGURE 18 In September 1981, a new division of Laidlaw, Total Recycling, began an experimental recycling project that covered one thousand households in Kitchener. Designed by Resource Integration Systems, a spinoff of the Is Five Foundation, the project assessed the effectiveness of various collection methods, including blue boxes. Originally scheduled for six months, it continued uninterrupted until 1983. Marking the launch of this initiative are (left to right) Eric Hellman, Steve Murray, Bill Kirkham, and Peter McGough. *Source:* Kitchener-Waterloo *Record* Photographic Negative Collection, 81-2107, Dana Porter Library, University of Waterloo.

down with ultraviolet light." Black was considered unattractive and the organizers did not want to go with the conventional green, so they settled on blue.[40]

Laidlaw's blue box program went citywide in Kitchener in 1983 and participation levels hit 85 percent almost immediately.[41] Ludolph recalls that they had little trouble engaging the public in the voluntary program. Bins containing educational information were left at the entrance of each home in the city. According to Ludolph: "When we distributed the 35,000 [blue boxes] I only had four people that said 'Come take this thing away, we're not going to do this.' I must tell you that within a week three of these people called back and said they had changed their mind."[42] Despite the popularity of the expanded program, in which Laidlaw had invested $500,000, it was nearly abandoned the following year when the company's contract with the city expired. Laidlaw attempted to recoup some of its costs in its follow-up bid, but Browning-Ferris Industries, a garbage

FIGURE 19   Total Recycling employee Bob LeDrew is seen stacking some of the 35,000 blue boxes set for distribution when the recycling program went citywide in Kitchener in 1983. *Source:* Kitchener-Waterloo *Record* Photographic Negative Collection, 83-1646, Dana Porter Library, University of Waterloo.

contractor without a recycling plan, submitted a bid that was $400,000 lower. However, at the ensuing General Council meeting, public support for the blue box program, combined with presentations from Ludolph, Paul Taylor of the Recycling Council of Ontario, Pollution Probe executive director Colin Isaacs, and a group of schoolchildren who recited a poem on the merits of recycling, persuaded the council to accept the higher bid.[43]

The blue box program continued to expand. In 1985, Laidlaw brought it to Mississauga and the Ontario Soft Drink Association made a deal with the provincial government: the Environmental Protection Act would be amended to allow the introduction of non-refillable but recyclable aluminum and plastic containers; in return, the association promised that it would recycle 50 percent of its containers by December 1988. To carry out its part of the bargain, it established Ontario Multi-Materials Recycling Inc. in 1986, which made an initial pledge of $1.5 million to expand the blue box program provincewide. Within the year, this pledge was increased to $20 million.[44]

In 2011 the blue box program serviced 95 percent of Ontario's population, diverting 904,850 tonnes of the province's waste from landfills. The spread of the blue box program highlights environmentalists' success in selling the concept of recycling to the public and the business community. Managed by Waste Diversion Ontario, which was created by an act of the provincial parliament, the program's costs are evenly distributed by the municipalities and Stewardship Ontario, a not-for-profit organization funded by the companies whose products are collected through the program. Furthermore, the program and its associated expertise have spread throughout Canada and internationally.[45]

On the other hand, this important victory for environmentalists obscures the fact that recycling is only one part of the solution to the waste problem. The complete solution, as outlined in Pollution Probe's 3Rs waste hierarchy, begins with a reduction of the throughput and continues with an emphasis on purchasing reusable goods. Whereas these actions demanded significant changes in the lifestyles of consumers, as well as major changes in the way producers operated, recycling was a relatively easy fix that enabled the public to feel good about itself without addressing the unsustainability of the modern consumer lifestyle. In this respect, it must be noted that the market share for soft drinks sold in refillable containers actually fell from 40 percent in 1986 to just 3 percent in 1993.[46]

RISE OF THE SECOND-WAVE ENGOs

The mid-1980s ushered in a period of renewed interest in the environment on the part of both government and the public. This second wave differed from the earlier environmental boom in noteworthy ways. At the outset of the first wave, environmental activists focused on localized pollution problems. Their focus expanded rapidly to include concern over the broader implications of energy and resource policies, but localized pollution issues continued to dominate. This changed in the 1980s as transnational concerns, such as acid rain, the depletion of the ozone layer, and the rapid decline of the planet's biodiversity, moved to the forefront.[47]

Coinciding with the new focus, the second wave of the environmental movement gave rise to a new type of activist organization. Whereas the movement had previously been dominated by localized groups, a new generation of pan-Canadian ENGOs emerged that were better equipped to raise the resources necessary to tackle emergent transnational problems and move into specific niches:

- The first of these pan-Canadian groups was Friends of the Earth Canada. The Canadian branch of Friends of the Earth International, it served as an umbrella organization that united existing Canadian ENGOs, with a particular emphasis on representing their interests in Ottawa. Founded in 1978, it gained prominence as interest in environmental concerns revived during the 1980s.[48]
- The Canadian Coalition on Acid Rain (CCAR) was formed in 1981 by twelve member groups that were concerned about the effects of acid rain on the Canadian environment and economy. In order to address the root cause, sulphur emissions from Canadian and American factories, the CCAR initiated educational campaigns as well as political lobbies in Ottawa and Washington, DC. It expanded to encompass a support base of fifty-eight organizations, representing over two million Canadians, but disbanded in 1991 after the Mulroney government in Canada and the Bush administration in the United States enacted the necessary clean air legislation.
- World Wildlife Fund Canada (WWFC), an offshoot of the Switzerland-based World Wildlife Fund, had been founded in 1967 by Senator Alan A. Macnaughton but was largely dormant until its incorporation as a legal foundation in 1982, with former Pollution Probe executive director Monte Hummel as its head. The WWFC's efforts to preserve wilderness areas and their natural inhabitants in Canada and abroad were aided by an annual budget of $4 million in 1988, a figure buoyed by the support of two trust funds and broad-based public support.
- Greenpeace, which continued to appeal to those who favoured direct action tactics, unified its Canadian operations under the Greenpeace Canada banner. A subsidiary of Greenpeace International, it opened its national headquarters in Toronto in 1987.
- Sierra Club Canada was established in 1989 as a grassroots collection of volunteer-driven provincial chapters. An independent outgrowth of the venerable US organization, the national body replaced unaffiliated provincial chapters in an effort to more effectively address its

concerns. Headed by activist-turned-environmental lawyer Elizabeth May, Sierra Club Canada quickly developed into the country's largest direct-membership ENGO.[49]

Not all second-wave ENGOs were inaugurated with a cross-Canada network of offices. The Canadian Environmental Defence Fund, formed to provide financial, technical, and organizational support to citizens engaged in environmental battles, began operations in 1984 with a single office in Toronto. The Pembina Institute was incorporated the following year, based out of an office in Alberta's Drayton Valley. Inspired by the 1982 Amoco Lodgepole Blowout, an incident that released sour gas into the air and resulted in the evacuation of local residents and farm animals, the organization focused on environmental issues related to the Canadian energy sector. The Sierra Legal Defence Fund and the David Suzuki Foundation were founded in Vancouver in 1990. Whereas the former was designed to pursue legal action to protect the Canadian environment – an objective differentiated it from the Ontario-centric CELA (1972) and the West Coast Environmental Law Association (1974) legal clinics by its national scope – the latter was founded as an environmental think tank that would capitalize on the celebrity of its eponymous co-founder. Despite each ENGO's association with a specific locale at the time of its founding, all four developed into important players in the Canadian environmental community, opening additional offices across the country.[50]

The second-wave ENGOs illustrate another important development in the Canadian environmental movement. During the first wave, very few in the Toronto community were able to make a career of their activism. With the notable exception of Energy Probe's Lawrence Solomon and Norm Rubin, activists rarely lasted more than a few years at their jobs. While some stayed within the general field – environmental consulting was popular among former Pollution Probe staffers – most were attracted to more stable careers in other industries. The second-wave organizations prided themselves on their professionalization, which opened up the possibility of long-term careers. This was highlighted in an interview with Monte Hummel published in 1990:

> I felt that in the interests of making progress on the issues it was legitimate to start building a few business principles into the organization. My views have only intensified in that respect. Now when you walk into my current offices ... you might as well be walking into the head office of IBM. Absolutely everybody has a computer at his or her station. At the WWF[C]

we take a business-like approach in what I think is the good sense of business, which is channelling your energy efficiently and accountably and getting results.[51]

Conclusion

The early 1980s were kind to Pollution Probe. A newfound focus on hazardous waste and personal health helped boost the organization's stagnant public profile, and it also enjoyed greater financial stability. This period was not without controversy, however, as demonstrated by executive director Colin Isaacs's resignation following the unfortunate decision to endorse "green" disposable diapers sold at Loblaws supermarkets. These events coincided with important developments in the local and national environmental community. Close to home, the breakaway Energy Probe Research Foundation expanded into international affairs, leading to the eventual creation of Probe International, while a collaboration between private and public interests resulted in the implementation of municipal recycling programs using the iconic, made-in-Ontario blue box. This period also witnessed the emergence of the second-wave ENGOs and a greater ability to address the transnational environmental issues then gaining prominence. Thus, even as Pollution Probe was enjoying renewed vigour, the terrain of the Canadian environmental movement was shifting considerably.

Just as Pollution Probe found itself at an interesting juncture at the outset of the second wave, such is the status of the history of the Canadian environmental movement. A small but growing field of study, the movement has received scant attention in the literature dealing with Canada in the 1960s and the ensuing explosion of social movements. There is no clear explanation for this, although it does appear that the environmental movement is inextricably linked to the 1970s – not the 1960s – in the minds of historians. Take, for example, the introduction to Campbell, Clément, and Kealey's *Debating Dissent* collection, which notes that "the founding of Greenpeace in Vancouver in 1971 symbolized the birth of the modern environmental movement."[52] As demonstrated in this book, there was a vibrant environmental movement across Canada before the advent of Greenpeace.

The exclusion of the environmental movement from the 1960s literature is a major oversight, as it was one of the most successful of the postwar movements. It is important to remind ourselves how much has changed since its emergence less than half a century ago. At that time, there were no federal or provincial ministries of the environment. Environmental education had not yet been incorporated in the school system. Few, in fact, were familiar with the words "environment" or "pollution." Since then, we have seen the creation of an extensive environmental regulatory regime. Environmental education has been integrated into the public school curriculum, and university Environmental Studies programs are the rule rather than the exception. "Environment" and "pollution" have become household words.

ENGOs such as Pollution Probe laboured to popularize concern for the environment and to bring about positive change. While it would be folly to credit any single ENGO with the sea change in public awareness of environmental issues, it is fair to say that Pollution Probe played an essential role. For many Canadians, particularly those living in Ontario during the late 1960s and early 1970s, Pollution Probe provided their first exposure to environmental activism. The organization's media-friendly stunts raised the profile of many issues. Pollution Probe also fostered the growth of the environmental movement at home in Toronto and beyond through the creation of additional organizations – CELA, CELRF, CAHE, COPE, Energy Probe, and its independent affiliates – and by sharing its expertise with other ENGOs.

This book also points to a bias in the study of postwar movements, particularly those emerging from the tumultuous 1960s. Existing studies highlight the leftist nature of those seeking social change. The assumption has been that environmentalists are inherently part of the political left. *The First Green Wave* challenges this widespread misconception, as Pollution Probe endeavoured to be politically nonpartisan, adopting a position of pragmatic centrism. It is true that some within the organization associated with the political left, just as others associated with the political right. Those responsible for making decisions, however, were unencumbered by political baggage and adopted a solutions-based approach to the issues. This, it turns out, was one of the keys to Pollution Probe's success.

Pollution Probe's pragmatism gave rise to two of its defining features: its ability to work with industry and its ability to work with government. The organization did not successfully tap into the general public as a revenue source until the second wave; consequently, funding from the business community was essential to its operations. As the history of Pollution Probe demonstrates, utilizing this funding without appearing to compromise its credibility was of the utmost importance. For years Pollution Probe's corporate funding went unchallenged because it held its donors to the same high standard it held others. It was only after that principle appeared to be compromised in the aforementioned Green Line fiasco that it found itself under scrutiny for its source of funding.

Collaboration with government was important because Pollution Probe – as well as CELA – played a role in dismantling the bipartite bargaining process. As early as 1971, Pollution Probe was invited to join representatives from government and industry as a recognized stakeholder in environmental issues. That year, Dr. Henry Regier joined the province-appointed Advisory Committee on Energy, which was formed to review Ontario's future energy needs and make recommendations to guarantee that they were met. In 1972 Gregory Bryce was appointed to the Toronto Recycling Action Committee, a subcommittee of the city's Department of Public Works, and Peter Love was appointed to the provincial Solid Waste Task Force and the accompanying Beverage Packaging and Milk Packaging working groups. In all three cases, Pollution Probe was the sole ENGO appointed to work alongside representatives from industry and, to a lesser extent, government. These appointments were made possible because from the outset Pollution Probe aimed to create a working relationship with the business community and politicians. In addition, Pollution Probe did an excellent job of establishing itself as an environmental stakeholder, presenting itself as a legitimate representative of the interests of the environment and the Ontario public, submitting carefully researched briefs to bodies such as the CRTC and Task Force Hydro, and providing a voice for those concerned about the state of the environment. When government was reluctant to investigate the apparent pesticide-induced death of numerous mallards, or the effect that adding a superstack to the Hearn Generating Station would have on air quality, Pollution Probe filled a role typically played by the government and brought together stakeholders at its own inquiries. It also branded itself effectively, using an advertising campaign to establish itself as the province's leading critic of industry and government on environmental issues. None of this would have mattered had Pollution Probe been a disreputable organization. As this book demonstrates, however, it was wholly reputable, not only for the reasons outlined above but also because it was affiliated from the outset with the Department of Zoology at the University of Toronto. With the support of departmental faculty, most notably Dr. Donald Chant, Pollution Probe could not be dismissed as a bunch of well-meaning but confused do-gooders.

Pollution Probe's invitation to the negotiating table did not mark a complete break with the bipartite bargaining paradigm. Just as progress was being made on this front, the mid-1970s saw a major dip in public and government interest in environmental issues. The decreased attention to environmental issues during this period – a natural part of the issue-attention cycle – meant that there was less pressure to include ENGOs in the making of deals. By the time popular interest in the environment revived among Canadians in the 1980s, a precedent had been established to allow ENGOs to participate in the process as stakeholders. This, combined with the greater sophistication of ENGOs during the 1980s, meant that environmentalists could no longer be excluded from such meetings.

It is worth noting that the movement is still young and continues to evolve. The groups examined here learned on the job how to run successful campaigns, raise funds, and organize effectively. It is also worth noting that the movement's pioneers were conspicuously white and from privileged, middle-class or better backgrounds. With the notable exception of Pollution Probe's Urban Team, little attention was devoted to the specific concerns of immigrant and working-class populations. This, of course, is consistent with environmental activism elsewhere in Canada and the United States during this period, and began to change only with the rise of the environmental justice movement in the 1980s. Studies of later periods will no doubt show greater diversity of backgrounds among Toronto's environmental activists.

The Toronto environmental community has grown considerably since the movement's origins. From only a handful through the 1970s, environmental organizations today number in the hundreds. Just as the localized groups of the first wave gave way to the pan-Canadian bodies of the second, more recent times have seen a proliferation of Toronto-centric organizations such as the Toronto Environmental Alliance and small, single-issue bodies such as Dads Against Dirty Air. In many ways, environmentalism is a thriving industry in the city – a response to humanity's penchant for ignoring the environmental consequences of their actions. As new threats are identified, it will continue to grow, built on the solid foundation laid by the movement's pioneers.

Afterword

RESEARCH FOR THIS BOOK began in earnest in May 2007, when I was granted access to Pollution Probe's internal archives. Having sold Ecology House at a handsome profit just a few years earlier, the organization was now operating out of the Traders Building, a mid-rise office building at 625 Church Street. On arriving for my initial visit, I was welcomed by Patty Chilton, the deputy executive director of Pollution Probe, and was given a tour of the premises. A far cry from the mismatched workplace of yore, the modern Pollution Probe headquarters more closely resembled those that emerged during the second wave, with neat offices and cubicles, a kitchenette, and a common area, all stocked with the latest equipment.

My tour ended in Pollution Probe's small library, which featured bookshelves full of the organization's many publications and a couple of filing cabinets full of newsletters, reports, and correspondence dating back to the ENGO's founding. It was at this point that I had a personal revelation. Sitting on the bookshelf was a copy of *The Canadian Junior Green Guide*. A primer on key environmental issues and concepts developed for Canadian youth, this heavily illustrated book was prepared by Teri Degler and Pollution Probe staff, with poetic contributions from literary icon Dennis Lee, and was published in 1990 by McClelland and Stewart. That same year, my parents gave me a copy for Christmas. The book was key in the development of my environmental ethos – both my grade six science fair project on acid rain and my grade seven project on composting were inspired by experiments detailed in its pages. Nonetheless, until I spotted

The Canadian Junior Green Guide in Pollution Probe's library, I had been completely unaware that the subject of my research had a deeply personal connection.

As an historian, I also took note of the state of historical memory in the venerable organization. Aside from Chilton, those I spoke with had little or no knowledge of Pollution Probe's colourful past, expressing surprise when I mentioned its pioneering and influential work. Tales of the group's early years, such as the Don River funeral and the dead ducks inquiry, were met with varying degrees of amusement and bewilderment. Likewise, with the exception of Donald Chant, key figures from the past were virtually unknown. This should come as no surprise. Pollution Probe, like its fellow ENGOs, is not a museum. Its employees have been hired for their ability to meet current needs and address contemporary concerns. This, coupled with a regular turnover of staff, has resulted in a void in the collective memory.

This is not to say that the organization does not, from time to time, endeavour to pay tribute to its early history. On 11 October 1990, a chestnut tree was planted outside the University of Toronto's Robarts Library in honour of twelve individuals credited as Pollution Probe's founders: Tony Barrett, Gregory Bryce, Sherry Brydson, Donald Chant, Merle Hanes, Monte Hummel, Brian Kelly, Ruth Kelly, Ann Love, Peter Middleton, Robert Mills, and Paul Tomlinson. While all made important contributions in the organization's early years, these names reveal the highly selective nature of choosing and recognizing founders. Given the role Brydson and Chant played in kick-starting Pollution Probe, their inclusion is obvious. Some of those named were key contributors who were among the hundreds who attended the group's founding meetings in spring 1969, while others joined in 1970 and 1971. Evidently, not all "founders" were part of the organization at the outset.

The criteria for recognition as a founder have been questioned by some of the very individuals cited. Ann Love joined Pollution Probe as a full-time staff member in the fall of 1970, a year and a half after it began operations. However, the former information and education coordinator acknowledged in an interview that she found her inclusion in the list somewhat puzzling. As she explained: "I was credited with being there earlier because I think they wanted to say that there were more women involved ... It was a politically correct move to call me a founder."[1] Whatever the rationale, her inclusion served to highlight the exclusion of her brother-in-law Peter Love, a key figure in the early years, whose involvement began in the spring of 1969. Pollution Probe lore was not limited to identifying alleged founding members. Peter Middleton spoke of the emergence of a "revisionist history" that focused on the contributions of a select few. Such a focus, he argued, downplays the significant role of the hundreds of volunteers and staff who enabled the organization to function in the early years.[2]

As with any exercise involving history and memory, each person has differing recollections of Pollution Probe's early days, its activities, and who the key contributors were. In the absence of an accurate historical record, a certain lore was bound to grow around Pollution Probe, typically emanating from those who had been involved with the organization over the years. Naturally, the select stories that emerged highlighted what these individuals considered to be most significant. Individuals and events that were singled out eventually found their way into print in newsletters or reports, solidifying their place in the "official" story of Pollution Probe. Malice was not necessarily involved, but, rather, natural human bias. Nonetheless, the consequences of selective storytelling have been very real. For instance, those who were present in the early years may feel slighted when their contributions remain outside the organization's lore. Some speculate that this may explain why the early core of this trailblazing organization is not particularly close-knit. Furthermore, the original diversity of its program seems to have been forgotten. Although Pollution Probe was founded in response to concerns over air pollution, it rapidly evolved to address such diverse issues as water and noise pollution, energy policy's effect on the environment, and urban planning's impact on the health of inner-city communities.

While the wild-eyed days of attention-grabbing activism appear to be a thing of the past, for Pollution Probe this had always been a means of bringing environmental issues to the forefront. Now under the leadership of CEO Bob Oliver and operating out of a North York office tower, the organization has restricted its focus in recent years to matters concerning clean air and water. While the narrower focus – necessary in order for the organization to define itself in the crowded ENGO landscape – is a far cry from the all-encompassing environmental focus of the past, Pollution Probe has remained, perhaps ironically given its lack of nostalgia, remarkably close to its founding principles. Results-driven, it continues to emphasize the importance of building partnerships among stakeholders. It remains politically nonpartisan and continues to work with the different levels of government. Likewise, the business community continues to be an important source of funding, as evidenced by Pollution Probe's stated goal of raising $1 million at the 2013 edition of its annual gala celebration, an industry-heavy event at which a table for eight costs $6,500 and sponsorship packages begin at $25,000.[3]

Since I began working on this book, I have had the opportunity to teach courses on the history of the environmental movement in Canada and the United States, Canadian environmental history, and Canada in the 1960s at the University of Prince Edward Island and Trent University. In each class, I would survey the students, asking how many had heard of Pollution Probe. One or two students might raise their hands. I would then ask how many had heard of Greenpeace. Inevitably, all of the students, or nearly all, would raise their hands. Both organizations were instrumental in shaping the environmental movement in Canada, yet only one has become a household name. Part of the explanation is that Greenpeace has maintained a more public persona as well as an international presence, while Pollution Probe has evolved into an organization that favours working behind the scenes from its Toronto headquarters.[4] At the same time, Greenpeace has been much more successful in propagating its early history through books, articles, and documentary films. There is power in history, and the failure to make its story known to the curious masses has been a disservice to the legacy of Pollution Probe. It is my hope that this book is just the beginning of the sharing of the story of Pollution Probe and its fellow ENGOs, and that scholars and former members alike will engage in this process.

Notes

Foreword

1 Yogi Berra, *The Yogi Book* (New York: Workman Publishing, 1998); "Wisest Fools," *The Economist*, 29 January 2005, http://www.economist.com/node/3598744.
2 *The Air of Death*, directed by Larry Gosnell (1967; Toronto: CBC Archive Sales, 2008), DVD.
3 Sherry Brydson, "Pollution: Is There a Future for Our Generation?" *The Varsity*, 24 February 1969, 1; Sherry Brydson, "Pollution Probe: Is Dunnville Dying?" *The Varsity*, 26 February 1969, 4-5; Sherry Brydson, "People May Be Dying in Dunnville: Why, Dr. Solandt?" *The Varsity*, 3 March 1969, 25; see also, Sarah Elton, "Green Power," *U of T Magazine* (Winter 1999), http://www.magazine.utoronto.ca/feature/canadian-environmental-movement/.
4 The gist of the story and a sense of the illustrations in *Captain Enviro: The Fight to Save the Maritimes!* (Committee of Environment Ministers and Council of Maritime Premiers, 1972) can be gained from: http://www.misterkitty.org/extras/stupidcovers/stupidcomics147.html. See also Mark J. McLaughlin, "Rise of the Eco-Comics: The State, Environmental Education and Canadian Comic Books, 1971-1975," *Material Culture Review/Revue de la culture matérielle* 77/78 (Spring/Fall 2013): 9-20. There were, of course, other stirrings of concern. Journalist Boyce Richardson published a thirty-two page pamphlet entitled *Pollution: Everybody's Business* with the Montreal *Star* in 1969. In British Columbia, the Society Promoting Environmental Conservation traces its beginnings to 1969. And as Owen Temby has shown, concerns about pollution were articulated at the municipal level in Ontario in the 1950s. Owen Temby, "Trouble in Smogville: The Politics of Toronto's Air Pollution during the 1950s," *Journal of Urban History* 39, 4 (July 2013): 669-89.
5 William Wordsworth, "The French Revolution as It Appeared to Enthusiasts at Its Commencement," *The Complete Poetical Works, by William Wordsworth ... With an introduction by John Morley* (London: Macmillan, 1888). Available at: http://www.bartleby.com/145/ww285.html.

6 For a nice demonstration of these points, see Brian Friel, "Faith Healer," in *Brian Friel: Plays One* (London: Faber and Faber, 1996), 327-76.
7 Wordsworth, "The French Revolution."
8 Berra, *The Yogi Book*.
9 On Greenpeace, see Frank Zelko, *Make It a Green Peace! The Rise of Countercultural Environmentalism* (Toronto: Oxford University Press, 2013), which includes reference to the wide range of other studies.
10 Ryan O'Connor, "Advertising the Environmental Movement: Vickers and Benson's Branding of Pollution Probe," *Rachel Carson Centre Perspectives* 1 (2013): 43-52, http://www.environmentandsociety.org/sites/default/files/seiten_aus_2013_i1_web_final2_kleiner-3_0.pdf adds an important dimension to this discussion, alluded to but not developed in the pages of this book.
11 Eric Hobsbawm, "Un historien et son temps present," in *Actes de la journée d'études de l'Institut d'Histoire du Temps Present* (Paris: CNRS, 1992), 98, cited by Mark Salber Phillips, "Rethinking Historical Distance: From Doctrine to Heuristic," *History and Theory* 50, 4 (December 2011): 12.
12 Mark Salber Phillips, "Reconsidering Historical Distance," the description of a doctoral training course that Phillips offered at Universiteit Gent in May 2011. A summary is available at: http://www.ugent.be/doctoralschools/en/doctoraltraining/courses/archive/2010-2011/2010-2011-reconsidering-historical-distance.htm.
13 Al Gore, *An Inconvenient Truth: The Planetary Emergency of Global Warming and What We Can Do about It* (Emmaus, PA: Rodale Press, 2006); Peter F. Sale, *Our Dying Planet: An Ecologist's View of the Crisis We Face* (Berkeley: University of California Press, 2011); James Gustave Speth, *Red Sky at Morning: America and the Crisis of the Global Environment* (New Haven, CT: Yale University Press, 2004); James Gustave Speth, *The Bridge at the End of the World: Capitalism, the Environment, and Crossing from Crisis to Sustainability* (New Haven, CT: Yale University Press, 2008).
14 Douglas Macdonald, *Business and Environmental Politics in Canada* (Peterborough, ON: Broadview Press, 2007), 120. See also George Hoberg, "North American Environmental Regulation," in ed. G. Bruce Doern and Stephen Wilks, *Changing Regulatory Institutions in Britain and North America* (Toronto: University of Toronto Press, 1998), 305-327; George Hoberg, "Governing the Environment: Comparing Canada and the United States," in ed. Keith G. Banting, George Hoberg, and Richard Simeon, *Degrees of Freedom: Canada and the United States in a Changing World* (Montreal and Kingston: McGill-Queen's University Press, 1997), 341-87; Mark S. Winfield, *Blue-Green Province: The Environment and the Political Economy of Ontario* (Vancouver: UBC Press, 2012).
15 Tzeporah Berman (with Mark Leiren-Young), *This Crazy Time: Living Our Environmental Challenge* (Toronto: Alfred A. Knopf Canada, 2011), 145.
16 Ola Tjornbo, Frances Westley, and Darcy Riddell (Social Innovation Generation@ University of Waterloo), Case Study 003. The Great Bear Rainforest Story (January 2010) available at: http://sig.uwaterloo.ca/highlight/case-study-the-great-bear-rainforest-story.
17 For more on this thread, see Graeme Wynn, "Rethinking Environmentalism," foreword to Justin Page, *Tracking the Great Bear: How Environmentalists Recreated British Columbia's Coastal Rainforest* (Vancouver: UBC Press, 2014) and this volume in toto.
18 Wordsworth, "The French Revolution"; Berra, *The Yogi Book*.

Introduction

1 Monte Hummel, interview with author, 23 January 2008, Toronto.
2 Groups from outside Toronto that were founded during this period include British Columbia's Scientific Pollution and Environmental Control Society (later renamed the Society Promoting Environmental Conservation; 1969) and Greenpeace (1971), Edmonton's Save Tomorrow Oppose Pollution (1970), the Saskatchewan Environmental Society (1970), Clear Hamilton of Pollution (since renamed Conserver Society of Hamilton and District; 1969), Montreal's *Societé pour vaincre la pollution* and the Society to Overcome Pollution (1970), and Halifax's Ecology Action Centre (1971). All of these groups, except for the Society to Overcome Pollution and Save Tomorrow Oppose Pollution, continue to operate.
3 Doug Macdonald, *The Politics of Pollution* (Toronto: McClelland and Stewart, 1991); G. Bruce Doern and Thomas Conway, *The Greening of Canada: Federal Institutions and Decisions* (Toronto: University of Toronto Press, 1994); Jennifer Read, "'Let Us Heed the Voice of Youth': Laundry Detergents, Phosphates and the Emergence of the Environmental Movement in Ontario," *Journal of the Canadian Historical Association* 7 (1996): 227-50; Arn Keeling, "Urban Waste Sinks as a Natural Resource: The Case of the Fraser River," *Urban History Review* 34, 1 (Fall 2005): 58-70; Mark J. McLaughlin, "Green Shoots: Aerial Insecticide Spraying and the Growth of Environmental Consciousness in New Brunswick, 1952-1973," *Acadiensis* 40, 1 (Winter/Spring 2011): 3-23; Philip Van Huizen, "Building a Green Dam: Environmental Modernism and the Canadian-American Libby Dam Project," *Pacific Historical Review* 79, 3 (2010): 418-53; Philip Van Huizen, "'Panic Park': Environmental Protest and the Politics of Parks in British Columbia's Skagit Valley," *BC Studies* 170 (Summer 2011): 67-92; Zelko, *Make It a Green Peace!*; Frank Zelko, "Making Greenpeace: The Development of Direct Action Environmentalism in British Columbia," *BC Studies* 142/143 (Summer/Autumn 2004): 197-239.
4 In his study of Canadians and their relationship with the environment, Neil S. Forkey singles out two environmental activist groups, Pollution Probe and Greenpeace, as fundamental in shaping the movement. Forkey, *Canadians and the Natural Environment to the Twenty-First Century* (Toronto: University of Toronto Press, 2012), 95.
5 Christopher J. Bosso, *Environment, Inc.: From Grassroots to Beltway* (Lawrence: University Press of Kansas, 2005), 45.
6 George M. Warecki, *Protecting Ontario's Wilderness: A History of Changing Ideas and Preservation Politics, 1927-1973* (New York: Peter Lang Publishing, 2000), 144.
7 These obstacles also confronted Canadian rights organizations and conservation groups. Dominique Clément, *Canada's Rights Revolution: Social Movements and Social Change, 1937-1982* (Vancouver: UBC Press, 2008), 15; Warecki, *Protecting Ontario's Wilderness*, 108.
8 Robert Paehlke, "Eco-History: Two Waves in the Evolution of Environmentalism," *Alternatives* 19, 1 (September 1992): 18-23. This periodization of the movement coincides with the "distinct waves of public concern for the environment in Ontario," as presented in Winfield, *Blue-Green Province*, 5, and the two waves discussed in Hoberg, "North American Environmental Regulation," 306.
9 John McCormick, *Reclaiming Paradise: The Global Environmental Movement* (Bloomington and Indianapolis: Indiana University Press, 1989), 47-48.
10 Samuel P. Hays, *Explorations in Environmental History: Essays* (Pittsburgh, PA: University of Pittsburgh Press, 1998), 380.

11 William K. Carroll, "Social Movements and Counterhegemony: Canadian Contexts and Social Theories," in *Organizing Dissent: Contemporary Social Movements in Theory and Practice*, 2nd ed., ed. William K. Carroll (Toronto: Garamond Press, 1997), 4.
12 Lara Campbell and Dominique Clément, "Introduction: Time, Age, Myth – Towards a History of the Sixties," in *Debating Dissent: Canada and the Sixties*, ed. Lara Campbell, Dominique Clément, and Gregory S. Kealey (Toronto: University of Toronto Press, 2012), 6.
13 The closest any of these chapters come to discussing the environmental movement is Catherine Carstairs, "Food, Fear, and the Environment in the Long Sixties," in ed. Campbell, Clément, and Kealey, *Debating Dissent*, 29-45. Other collections dedicated to the history of Canada in the 1960s are Dimitry Anastakis, ed., *The Sixties: Passion, Politics, and Style* (Montreal and Kingston: McGill-Queen's University Press, 2008), and M. Athena Palaeologu, ed., *The Sixties in Canada: A Turbulent and Creative Decade* (Montreal: Black Rose Books, 2009). While not specifically Canadian in scope, Karen Dubinsky et al., eds., *New World Coming: The Sixties and the Shaping of Global Consciousness* (Toronto: Between the Lines, 2009), does contain a number of essays about Canada during the 1960s.
14 Myrna Kostash, *Long Way from Home: The Story of the Sixties Generation in Canada* (Toronto: James Lorimer, 1980); Doug Owram, *Born at the Right Time: A History of the Baby Boom Generation* (Toronto: University of Toronto Press, 1996); Bryan Palmer, *Canada's 1960s: The Ironies of Identity in a Rebellious Era* (Toronto: University of Toronto Press, 2009). Specialized studies include Clément, *Canada's Rights Revolution*; John Hagan, *Northern Passage: American Vietnam War Resisters in Canada* (Cambridge, MA: Harvard University Press, 2001); Sean Mills, *The Empire Within: Postcolonial Thought and Political Activism in Sixties Montreal* (Montreal and Kingston: McGill-Queen's University Press, 2010); Cyril Levitt, *Children of Privilege: Student Revolt in the Sixties* (Toronto: University of Toronto Press, 1984); Gary Kinsman and Patrizia Gentile, *The Canadian War on Queers: National Security as Sexual Regulation* (Vancouver: UBC Press, 2009); Stuart Henderson, *Making the Scene: Yorkville and Hip Toronto in the 1960s* (Toronto: University of Toronto Press, 2011); Roberta Lexier, "The Canadian Student Movement in the Sixties: Three Case Studies," PhD dissertation, University of Alberta, 2009.
15 Campbell and Clément, "Introduction," 6.
16 Martin L. Friedland, *The University of Toronto: A History* (Toronto: University of Toronto Press, 2002), 525-42; Owram, *Born at the Right Time*, 240, 295-96; Palmer, *Canada's 1960s*, 284; Margaret Webb, "The Age of Dissent," *U of T Magazine* (Spring 2002), http://www.magazine.utoronto.ca/feature/the-age-of-dissent/.
17 Zelko, "Making Greenpeace."
18 Macdonald, *Business and Environmental Politics in Canada*, 120.
19 Hoberg, "North American Environmental Regulation," 309.
20 Ibid.; Hoberg, "Governing the Environment, 352; Macdonald, *Business and Environmental Politics in Canada*, 120.
21 Winfield, *Blue-Green Province*, 28-30.
22 Alan Frizzell and John H. Pammett, eds., *Shades of Green: Environmental Attitudes in Canada and around the World* (Montreal and Kingston: McGill-Queen's University Press, 1997); Judith I. McKenzie, *Environmental Politics in Canada: Managing the Commons into the Twenty-First Century* (Toronto: Oxford University Press, 2002), 89-99; "Environment a Priority for More Canadians, Poll Suggests," CBC News, 8 November 2006, http://www.cbc.ca/news/canada/story/2006/11/08/environment-poll.html; "The Canada's World Poll,"

Environics Research Group and the Simons Foundation, January 2008, http://www.environicsinstitute.org/PDF-CanadasWorld.pdf; "Canadians Continue to Voice Strong Support for Actions to Address Climate Change, Including an International Treaty and Carbon Taxes," Environics Research Group, 1 December 2011, http://www.environics.ca/reference-library?news_id=109.
23 "Canada Pulls Out of Kyoto Protocol," CBC News, 12 December 2011, http://www.cbc.ca/news/politics/story/2011/12/12/pol-kent-kyoto-pullout.html; Meagan Fitzpatrick, "Environment Canada Job Cuts Raise Concerns," CBC News, 4 August 2011, http://www.cbc.ca/news/politics/story/2011/08/04/pol-environment-job-cuts.html; Laura Payton, "Radicals Working against Oilsands, Ottawa Says," CBC News, 9 January 2012, http://www.cbc.ca/news/politics/story/2012/01/09/pol-joe-oliver-radical-groups.html.

CHAPTER 1: THE AIR OF DEATH AND THE ORIGINS
OF TORONTO'S ENVIRONMENTAL ACTIVIST COMMUNITY

1 Bosso, *Environment, Inc.*, 18.
2 "Larry Gosnell – Biography," 21 June 1972, Biography A-Z 1974-98, Canadian Broadcasting Corporation Reference Library (CBCRL).
3 Marc St-Pierre, "Footprints: Environment and the Way We Live," National Film Board, n.d., http://www3.nfb.ca/footprints/nfb-and-environment/the-early-years.html?part=3.
4 D.B. Jones, *Movies and Memoranda: An Interpretative History of the National Film Board of Canada* (Ottawa: The Canadian Film Institute, 1981), 82.
5 St-Pierre, "Footprints." According to Gosnell's widow, Denise, the decision to pull the documentary was made by executives at the NFB due to industry pressure. Larry was not notified of the decision. Denise Gosnell, interview with author, 19 March 2008, Toronto.
6 Canadian Radio-Television Commission (CRTC), *Public Hearing, In Connection with the Preparation, Production and Broadcasting of the CBC Television Programme entitled 'Air of Death'* [transcript of hearing] (Ottawa: L.A. Gillespie and Associates, 1969), 58, Larry Gosnell papers (LGP), in the possession of Denise Gosnell.
7 Ralph Thomas, "So Choose Sides: Earl or Stanley," Toronto *Star*, 12 November 1966, 28.
8 Murray Creed, interview with author, 28 January 2008, conducted by telephone; Untitled timeline, 20 February 1969, LGP.
9 Untitled timeline, 20 February 1969, LGP. For a scholarly treatment of earlier air pollution problems in Toronto, see Temby, "Trouble in Smogville," 669-89.
10 K.C. Walton, "Environmental Fluoride and Fluorosis in Mammals," *Mammal Review* 18, 2 (June 1988): 83; Transcript of discussion, Jim McLean, George Salverson, and Larry Gosnell, n.d., 1, LGP; "Dunnville Pollution Investigation," n.d., 3, LGP.
11 Transcript of discussion, Jim McLean, George Salverson, and Larry Gosnell, n.d., 4, LGP.
12 Gary Dunford, "Farmer's Diary Tells the Story of Six-Year Pollution Fight," Toronto *Star*, 30 October 1967, 31.
13 J.S. Cram, "Downwind from Disaster," *Family Herald*, 26 October 1967, 12-15; Pollution Inquiry Committee, *Report of the Committee Appointed to Inquire into and Report upon the Pollution of Air, Soil, and Water in the Townships of Dunn, Moulton, and Sherbrooke, Haldimand County* (Toronto: Queen's Printer, 1968), 346.
14 Transcript of discussion, Jim McLean, George Salverson, and Larry Gosnell, n.d., 23, LGP.

15 Since Waldbott was not licensed to practice medicine in Ontario, it would have been illegal for him to conduct physical examinations. CRTC, *Public Hearing*, 370-71; *The Air of Death*, Gosnell, DVD.
16 Albert W. Burgstahler, "George L. Waldbott – A Pre-Eminent Leader in Fluoride Research," *Fluoride* 31, 1 (February 1998): 2-4; Ontario Royal Commission on Fluoridation, *Report of the Committee Appointed to Inquire into and Report upon Fluoridation of Municipal Water Supplies* (Toronto: Queen's Printer, 1961); Catherine Carstairs and Rachel Elder, "Expertise, Health, and Popular Opinion: Debating Water Fluoridation, 1945-1980," *Canadian Historical Review* 89, 3 (September 2008): 348.
17 CRTC, *Public Hearing*, 363.
18 "Air Pollution," *Matinee*, CBC Radio, 19 October 1966, produced by Rodger Schwass, CBCRL.
19 CRTC, *Public Hearing*, 207-10.
20 Omond Solandt to Hugh McMahon, lawyer for ERCO, 18 March 1969, Omond M. Solandt fonds, B93-0041/038, University of Toronto Archives (UTA).
21 *The Air of Death*, Gosnell, DVD.
22 "The Audience and Its Reactions to a CBC-TV Documentary 'Special' on Air Pollution," CBC Research Department, December 1967, 3, Canadian Broadcasting Corporation fonds, RG 41, vol. 571, file no. 70, Library and Archives Canada (LAC).
23 Emphasis in original. Arthur Laird to Murray Creed, memo, 16 January 1968, RG 41, vol. 571, file no. 70, LAC.
24 Roy Shields, "TV Tonight," Toronto *Star*, 23 October 1967, 28.
25 Bob Blackburn, "In Blackburn's View," Toronto *Telegram*, 23 October 1967, 44.
26 "Doctor Says Two Struck by Fluorosis," *Globe and Mail*, 20 October 1967, 29.
27 Quoted in Terrance Wills, "Province Orders Fluorosis Probe around Dunnville," *Globe and Mail*, 28 October 1967, 1-2.
28 Omond Solandt to Sir Owen Wansbrough-Jones, 1 November 1967, Omond M. Solandt fonds, B93-0041/038, UTA.
29 Sir Owen Wansbrough-Jones to Omond Solandt, 26 October 1967, Omond M. Solandt fonds, B93-0041/038, UTA.
30 Pollution Inquiry Committee, *Report of the Committee*, xiii-xv; Solandt to Wansbrough-Jones, 1 November 1967, Omond M. Solandt fonds, B93-0041/038, UTA.
31 Terry Tremayne, "Fluorides Affect 2 More Victims, Doctor Asserts," *Globe and Mail*, 13 November 1967, 1-2; Carstairs and Elder, "Expertise, Health, and Popular Opinion," 353; "Personal Mention," *Industrial Canada* [Official publication of the Canadian Manufacturers Association], January 1968, 53; "Pollution Inquiry Rigged Claims MLA," Hamilton *Spectator*, 1 March 1969, 4.
32 Quoted in "Don't Believe the Cranks on Air Pollution – UK Expert," Toronto *Star*, 7 December 1967, 66. Lawther's comments drew criticism from Alderman Tony O'Donohue, who referred to Lawther as "the most biased man in the world in his intent to throw air pollution control protagonists into disrepute." "Dymond Influenced, O'Donohue Charges," *Globe and Mail*, 8 December 1967, 5.
33 Pollution Inquiry Committee, *Report of the Committee*, 12.
34 This point is made in Ella Haley, "Methodology to Deconstruct Environmental Inquiries Using the Hall Commission as a Case Study," PhD dissertation, York University, 2000, 211-14.

35 "Plant Closure Unnecessary," Toronto *Telegram*, 28 October 1967, 4; CRTC, *Public Hearing*, 120, 135-41; *Ontario Committee of Inquiry on Allegations Concerning Pollution in the Townships of Dunn, Moulton, and Sherbrooke* [transcript of hearing] (Toronto: Nethercut and Young, 1968), 136-40, 608-12, 645-46, Records of the Committee Appointed to Inquire into and Report upon the Pollution of Air, Soil and Water in the Township of Dunn, Moulton and Sherbrooke, Haldimand County, RG 18-155, AO; Haley, "Methodology to Deconstruct Environmental Inquiries," 260-62.

36 J.P. Gilmore, Vice-President Planning and Assistant Chief Operating Officer, to CBC executives, 24 January 1968, LGP. As Gilmore pointed out, *"The Corporation must take a firm position in support of the program work done for this project. We must support the people who did that work and who saw to its production."*

37 Marcel Munro to Max E. Weissengruber [Secretary of Committee of Enquiry], 22 March 1968, Canadian Broadcasting Corporation fonds, RG 41, vol. 571, file no. 70, LAC.

38 Quoted in Pollution Inquiry Committee, *Report of the Committee*, 347; George Waldbott, "Tried to Testify on Fluoride," Toronto *Star*, 14 May 1968, 6; CRTC, *Public Hearing*, 432-36.

39 Quoted in "Dunnville Probe Ignored Him Detroit Fluoride Man Claims," Toronto *Star*, 3 May 1968, 3.

40 Pollution Inquiry Committee, *Report of the Committee*, 347.

41 Ibid., 285-86, 296, 302, 307.

42 Omond Solandt to G.E. Hall, 21 January 1969, Omond M. Solandt fonds, B93-0041/038, UTA.

43 Gavin Henderson, "Air of Death," *Globe and Mail*, 27 February 1969, 6.

44 J.M. Anderson to Larry Gosnell, 20 January 1969, LGP.

45 Henry Regier to Larry Gosnell, 16 January 1969, LGP.

46 Donald Chant to Stanley Burke, 23 December 1968, LGP.

47 F.K. Foster, CRTC Secretary, "Public Announcement," 18 December 1968, LGP; F.K. Foster, CRTC Secretary, *"Notice of Public Hearing,"* 4 February 1969, LGP.

48 "CRTC 'Air' Hearing Right Is Asserted," Toronto *Telegram*, 18 March 1969, 8.

49 Murray Creed to E.S. Hallman, 17 January 1969, Canadian Broadcasting Corporation fonds, RG 41 vol. 571, file no. 70, LAC.

50 Victor Yannacone and Larry Gosnell, transcript of telephone conversation, 1 January 1969 (AM), LGP.

51 CRTC, *Public Hearing*, 6.

52 Ibid., 301.

53 Ibid., 430-31.

54 Ibid., 504.

55 George F. Davidson to Larry Gosnell, 31 March 1969, LGP. Praise for Gosnell was echoed in the *Globe and Mail* coverage of the CRTC hearing, which noted that "Gosnell himself radiated patience and integrity throughout the hearing, and it was hard to think of him as a reckless or careless producer; the image just didn't fit." Leslie Millin, "Air of courtroom about Air of Death," *Globe and Mail*, 22 March 1969, 31.

56 E.S. Hallman to Larry Gosnell, 1 April 1969, LGP.

57 Murray Creed, interview with author, 28 January 2008, conducted by telephone; "Producer Takes Blame at 'Air' Inquiry," Toronto *Telegram*, 19 March 1969, 3; George F. Davidson to Larry Gosnell, 31 March 1969, LGP.

58 CRTC, *Public Hearing*, 4, 7-9.

59 Murray Creed, memo to CBC Agriculture and Resources employees, 15 July 1970, LGP.
60 This point is made in Peter G. Cook and Myles A. Ruggles, "Balance and Freedom of Speech: Challenge for Canadian Broadcasting," *Canadian Journal of Communication* 17, 1 (1992), http://www.cjc-online.ca/index.php/journal/article/viewArticle/647/553#. The role of *The Air of Death* in this process is also acknowledged in Diane Burgess, "*Kanehsatake* on *Witness:* The Evolution of CBC Balance Policy," *Canadian Journal of Communication* 25, 2 (2000), http://www.cjc-online.ca/index.php/journal/article/ viewArticle/1153/1072.
61 Stanley Burke was credited as a co-founder of this group in contemporary media reports, but he did not play an active role. James Bacque, interview with author, 24 March 2008, Toronto. "Pollution Teach-in," Toronto *Telegram*, 11 November 1967, 9; "Pollution Group Plans 'Breathe-in,'" Toronto *Star*, 14 November 1967, 23.
62 Bacque, ibid.
63 "Pollution Teach-in," 9.
64 "Proposed Pollution Conference," 25 July 1967, Ontario Pollution Control, Correspondence of the Deputy Minister of Health, RG 10-06 b397312, Archives of Ontario (AO). As the date of this press release indicates, the conference's announcement predated the broadcast of *The Air of Death* by approximately three months.
65 "No Chance of Pollution-Free Society, Simonett Tells Delegates," *Globe and Mail*, 5 December 1967, 4; "GASP in the Street," Toronto *Telegram*, 7 December 1967, 41; Mack Laing, "Cough-and-Go for GASP," Toronto *Telegram*, 9 December 1967, 8; "Easter Breathe-in to Protest Pollution," Toronto *Star*, 9 December 1967, 37.
66 "Don't Believe the Cranks on Air Pollution – UK Expert," 66; "Crank Label or No, GASPer Sees Victory," *Globe and Mail*, 9 December 1967, 5; Laing, "Cough-and-Go for GASP," 8.
67 "Crank Label or No, GASPer Sees Victory," 5; "Easter Breathe-in to Protest Pollution," 37.
68 Ibid.; James Bacque, interview with author, 24 March 2008, Toronto.
69 "Pollution Fighters Demand Disclosure of Full Medical Facts in Fluoride Probe," Toronto *Star*, 26 January 1968, 29.
70 Charter, Group Action to Stop Pollution, 26 January 1968, Tony O'Donohue fonds, series 1250, file 684, City of Toronto Archives.
71 "Easter Breathe-in to Protest Pollution," 37; James Bacque, interview with author, 24 March 2008, Toronto. According to Denise Gosnell, her husband had no involvement with the group. Stanley Burke also denies any involvement, aside from chairing the 8 December 1967 meeting. Denise Gosnell, interview with author, 19 March 2008, Toronto; Stanley Burke, interview with author, 25 January 2008, Stella, ON.
72 "O'Donohue Heads Up Clean Air Campaign," Toronto *Star*, 31 January 1969, 50. It turns out that Bacque was unaware that O'Donohue had resurrected the group in his absence. He assumed that the group had died when he left for France; by the time he returned to Toronto, GASP had fallen dormant for the second time. Bacque noted during our interview on 24 March 2008 that this was the first he had heard of the organization's continuation in his absence.
73 Dr. Alfred Bernhart and Tony O'Donohue (GASP), brief to CRTC, 6, 4 March 1969, LGP.
74 Brydson, "Pollution," 1. A second article published in the same edition discussed the study week findings. As was noted, "Our findings paint a picture of death and destruction unless our generation, the heirs of a poisoned earth, get together and persuade the public, private industry and the government to do something about it." It then invited readers to join "the U of T Pollution Patrol." "Pollution Patrol," *The Varsity*, 24 February 1969, 4.

75 Bolded text in original. Brydson, "Pollution Probe," 4-5; Brydson, "People May Be Dying in Dunnville," 25.
76 Friedland, *The University of Toronto*, 525-42; Claude Bissell, *Halfway Up Parnassus: A Personal Account of the University of Toronto, 1932-1971* (Toronto: University of Toronto Press, 1974), 122-59.
77 Stanley Zlotkin, interview with author, 19 February 2008, Toronto.
78 Brian Kelly, interview with author, 12 January 2009, conducted by telephone.
79 Donald Chant, interview with author, 18 November 2007, conducted by telephone; *A Celebration of the Life of Dr. Donald Chant* (booklet published by family, 2008), 3.
80 Donald Chant, "CODA," unpublished memoirs, Donald Chant papers, in the possession of Merle Chant.
81 Rob Mills, "Letter from the President," *Probe Newsletter* 1, 2 (1 April 1969): 5, Pollution Probe papers (PPP), located at the Pollution Probe headquarters.
82 John Ayre, "A Student Puts the Case for Reform as Our New Campus Activists See It," University of Toronto *Graduate* 3, 4 (January 1971): 16; Chris Plowright, interview with author, 30 January 2008, conducted by telephone; Ralph Brinkhurst, interview with author, 8 August 2008, conducted by telephone; Henry Regier, interview with author, 14 August 2008, conducted by telephone.
83 Pollution Probe, brief to CRTC, 5 March 1969, "'Air of Death' Pollution Probe Brief to CRTC 1969," PPP. A brief was also submitted by Michael Kesterton, features editor at *The Varsity*, regarding a viewing of *The Air of Death* hosted by the campus newspaper. As he explained, the crowd of 242 was overwhelmingly in support of the CBC program. Michael Kesterton, brief to CRTC, 5 March 1969, LGP.
84 Pollution Probe was represented by Brydson, who provided testimony on 20 March 1969. CRTC, *Public Hearing*, 486-98.

Chapter 2: The Emergence of Pollution Probe

1 "Public Inquiry into the Death of Ducks on Ward's Island: Recommendations," 8 July 1969, Duck Inquiry – Correspondence, Toronto Island – Pesticides 1969, F1058 MU7338, AO.
2 "Probe Will Ask Writ against Insecticide," *Globe and Mail*, 19 July 1969, 5; Untitled article, *Probe Newsletter* 1, 5 (19 September 1969): 1, PPP.
3 "Elections," *Probe Newsletter* 1, 1 (8 March 1969): 3, PPP; Elton, "Green Power"; Donald Chant, interview with author, 18 November 2007, conducted by telephone.
4 John Coombs, interview with author, 13 November 2008, conducted by telephone.
5 "Meeting: Monday, March 17," *Probe Newsletter* 1, 2 (1 April 1969): 1, PPP.
6 Farley Mowat, *Rescue the Earth! Conversations with the Green Crusaders* (Toronto: McClelland and Stewart, 1990), 33.
7 This was a matter addressed in many interviews. Monte Hummel, interview with author, 23 January 2008, Toronto; Stanley Zlotkin, interview with author, 19 February 2008, Toronto; Peter Middleton, interview with author, 21 February 2008, Toronto; Rob Mills, interview with author, 25 September 2008, conducted by telephone; Brian Kelly, interview with author, 12 January 2009, conducted by telephone.
8 Lynn Spink, interview with author, 12 March 2008, Toronto. The important role of summer camp was a common theme in interviews. It was raised by Middleton, Hummel, and

Mills, ibid., and by Gregory Bryce, interview with author, 18 March 2008, conducted by telephone.
9 Sharon Wall, *The Nurture of Nature: Childhood, Antimodernism, and Ontario Summer Camps, 1920-55* (Vancouver: UBC Press, 2009), 20.
10 Rob Mills, interview with author, 25 September 2008, conducted by telephone.
11 Ralph Brinkhurst, interview with author, 8 August 2008, conducted by telephone.
12 Ann Rounthwaite, interview with author, 10 July 2008, conducted by telephone.
13 Rob Mills, interview with author, 25 September 2008, conducted by telephone.
14 John Coombs, interview with author, 13 November 2008, conducted by telephone.
15 Peter Middleton, interview with author, 21 February 2008, Toronto.
16 Tony Barrett, "And from the Administration," *Probe Newsletter* 1, 2 (1 April 1969): 3, PPP.
17 Brian Kelly, "How to Form Your Own Pollution Probe Group," 22, Projects/Reports/Submissions 1970s, PPP; Donald Chant, interview with author, 18 November 2007, conducted by telephone; Peter Middleton, interview with author, 21 February 2008, Toronto; Ralph Brinkhurst, interview with author, 8 August 2008, conducted by telephone; Henry Regier, interview with author, 14 August 2008, conducted by telephone.
18 Brian Kelly, interview with author, 12 January 2009, conducted by telephone.
19 Donald Chant, interview with author, 18 November 2007, conducted by telephone; Monte Hummel, interview with author, 23 January 2008, Toronto.
20 Claude Bissell, "Remarks to Pollution Probe Meeting on May 27th, 1970, at the St. Lawrence Centre," Office of the President, A77-0019/024, UTA.
21 Friedland, *The University of Toronto*, 526-35; Donald Chant, interview with author, 18 November 2007, conducted by telephone.
22 Brian Kelly, interview with author, 12 January 2009, conducted by telephone.
23 "Probe of Duck Deaths Told of Experiment," *Globe and Mail*, 10 December 1969, 5; Martin H. Edwards, *Did Pesticides Kill Ducks on Toronto Island? Report of the Royal Commission Appointed to Inquire into the Use of Pesticides and the Death of Waterfowl on Toronto Island* (Toronto: Queen's Printer, 1970), 1.
24 "Ex-Parks Man Tells Duck Probe He Faked Diazinon Story," Toronto *Star*, 12 December 1969, 35.
25 Michael Fitzgerald, "Dead Ducks Don't Faze Tommy," Toronto *Telegram*, 4 June 1969, 41.
26 "Island Ducks 'Set Pesticide Record,'" Toronto *Telegram*, 7 July 1969, 1.
27 "Insecticide Blamed in Duck Deaths," Toronto *Star*, 26 June 1969, 39.
28 "Parks Chief's Evidence Contradicts TV Interview," Toronto *Star*, 8 July 1969, 17; "Public Inquiry into the Death of Ducks on Ward's Island: Recommendations," 8 July 1969, Duck Inquiry – Correspondence, Toronto Island – Pesticides 1969, F1058 MU7338, AO.
29 "Allen Says Ducks Inquest 'Witch Hunt,'" Toronto *Star*, 16 July 1969, 41; "Pesticide That Killed Ducks Used on Island Today," Toronto *Star*, 18 July 1969, 1; "Court Injunction to Be Sought to Stop Island Spraying," Toronto *Star*, 19 July 1969, 12; "No Poison Spray on Island Pollution Group Urges," Toronto *Star*, 21 July 1969, 3; "Dymond Asks Metro to Stop Using Insecticide until Probe," Toronto *Star*, 24 July 1969, 3.
30 "Annual Report of the Pesticides Advisory Board [1969]," 5, Pesticides Advisory Board, RG 12-48, b352562, AO.
31 Edwards, *Did Pesticides Kill Ducks on Toronto Island?* 8.
32 Monte Hummel, interview with author, 23 January 2008, Toronto.
33 "Anti-Litter Campaign," *Probe Newsletter* 1, 5 (19 September 1969): 3, PPP.

34 The refuse was promptly cleaned up following the event. "Group Plans to Nab Road Litterers," *Globe and Mail*, 20 August 1969, 5.
35 Untitled press release, 19 August 1969, Pollution – General 1969, F1058 MU7338, AO.
36 "No Place for Amateur Policemen," *Globe and Mail*, 22 August 1969, 6; "Amateur Policemen," *Globe and Mail*, 26 August 1969, 6.
37 "Can't Enforce Litter Laws Police Admit," Toronto *Star*, 26 August 1969, 1; "Anti-Litter Campaign," *Probe Newsletter* 1, 5 (19 September 1969): 3, PPP.
38 Peter Taylor and Clive Attwater, "Probe Report on Soft Drink Containers," *Probe Newsletter* 2, 6 (October 1970): 5, PPP.
39 Untitled press release, 19 August 1969, Probe – Misc. Press Releases, F1058 MU7330, AO.
40 Ken McGowen to J.R. Thomas, 9 June 1970, Soft Drinks Containers – Correspondence, F1058 MU7336, AO.
41 "Kerr to Press for 5-Cent Bottle Deposits," Toronto *Star*, 21 May 1970, 77.
42 Terry O'Malley, interview with author, 8 July 2008, conducted by telephone.
43 Peter Middleton, interview with author, 21 February 2008, Toronto.
44 Rob Mills, interview with author, 25 September 2008, conducted by telephone.
45 Maggie Siggins, *Bassett* (Toronto: James Lorimer, 1979), 173.
46 For an analysis of this campaign, see O'Connor, "Advertising the Environmental Movement."
47 Thomas Claridge, "Pollution Probe Mourns for Beloved, Dead Don," *Globe and Mail*, 17 November 1969, 1.
48 Ibid.; "Don River Dead," *The Varsity*, 12 November 1969, 10; "Mock Rites Mourn Death of Don River Killed by Pollution," Toronto *Star*, 17 November 1969, 21; Rob Mills, interview with author, 25 September 2008, conducted by telephone; John Coombs, interview with author, 13 November 2008, conducted by telephone. The "hearse" utilized in the funeral procession was actually Peter Love's green station wagon. Ray Ford, "Death and rebirth of the Don River," *Canadian Geographic* (June 2011), http://www.canadiangeographic.ca/magazine/jun11/don_river_watershed.asp.
49 Quoted in Michael William Doyle, "Staging the Revolution: Guerrilla Theatre as a Countercultural Practice, 1965-1985," in *Imagine Nation: The American Counterculture of the 1960s and '70s*, ed. Peter Braunstein and Michael William Doyle (New York: Routledge, 2002), 74.
50 Meredith Ware, interview with author, 2 December 2008, conducted by telephone.
51 "Stanfield Sees Don River, Says 'It's a Mess,'" Toronto *Star*, 14 August 1970, 1; "150 Skip Classes, Study Pollution," Toronto *Star*, 9 April 1970, 63.
52 Mack Laing, "Power Plant's Evil Waste," Toronto *Telegram*, 3 November 1967, 10.
53 "City Told Giant Stack Will Help Cut Pollution," *Globe and Mail*, 23 September 1969, 10.
54 Thomas A. Beckett, "Air Pollution," *Globe and Mail*, 28 August 1969, 6.
55 Ross H. Hall to D.A. Chant, 24 September 1969, Hearn Generating Station Hydro – Inquiry 1970, F1058 MU7339, AO.
56 D.A. Chant to R.H. Hall, 31 September 1969, Hearn Generating Station Hydro – Inquiry 1970, F1058 MU7339, AO.
57 Untitled press release, 22 October 1969, Hearn Generating Station Hydro – Inquiry 1970, F1058 MU7339, AO.
58 "The Ontario Hydro Is Getting Ready to Give It to You from Great Heights," Toronto *Telegram*, 27 October 1969, 54. This advertisement ran three days earlier in *The Varsity*.
59 "Toronto Group Plans Probe of 700-Foot Hydro Stack," Toronto *Star*, 27 October 1969, 34.

60 "Surgeon Links Air Pollution to Rise in 'Respiratory Cripples,'" Toronto *Star*, 23 February 1970, 1.
61 "Remarks by George Gathercole, Chairman, Ontario Hydro, to the Inquiry into Air Pollution," Probe vs Ont. Hydro 1973, 24 February 1970, 9, F1057 MU7352, AO.
62 "Hydro Says 700-ft. Stack Best Way to Clean Air," Toronto *Star*, 24 February 1970, 9.
63 "Commission on Air Pollution in Metropolitan Toronto," 23-24 February 1970, Hearn Generating Station Hydro – Inquiry 1970, F1058 MU7339, AO.
64 "Hearn to Switch to Gas, See Rise in Hydro Rates," *Globe and Mail*, 30 June 1970, 5.
65 Brian Kelly, interview with author, 12 January 2009, conducted by telephone.
66 Read, "'Let Us Heed the Voice of Youth.'"
67 "Dishing the Dirt on Phosphates," CBC Digital Archives, originally broadcast on 8 February 1970, http://archives.cbc.ca/environment/pollution/topics/1390/.
68 "Detergents," *Probe Newsletter* 2, 2 (31 March 1970): 3-6, PPP.
69 Donald Chant, "A Brief to Prime Minister John P. Robarts," *Probe Newsletter* 2, 3 (8 April 1970): 3-5, PPP.
70 Read, "'Let Us Heed the Voice of Youth,'" 249. The success of this campaign has resulted in the incorrect belief that Pollution Probe "first emerged as a professor-student action group focused on Great Lakes pollution." Forkey, *Canadians and the Natural Environment to the Twenty-First Century*, 96.
71 "How Mrs. Boston Burst a Great Detergent Bubble," *Maclean's*, April 1970, 3.
72 "Students Exhorted to Fight Pollution," *Globe and Mail*, 1 November 1969, 5; "A Statement of Sorts," n.d., Press release, PPP.
73 Michael Enright, "PM Was Warmer This Year," *Globe and Mail*, 5 March 1970, 1. Ralph Brinkhurst told the story of a rather amusing exchange between Trudeau and Tony Barrett, who was wearing his "Do it" button. As he recalls, "Trudeau looked ... at this young man, sort of cocked his eyebrow and said, with obvious innuendo, 'Oh, what does that mean?' And ... [Barrett] instantly replied, 'It means think clean, sir.' *[laughs]* We were dealing with some smart kids. It really took him [Trudeau] aback." Whether this occurred at the same event covered in the *Globe and Mail* article is unclear. Ralph Brinkhurst, interview with author, 8 August 2008, conducted by telephone.
74 David Crane, "Tory Talks to Protesters," *Globe and Mail*, 11 May 1970, 5.
75 Rob Mills, "Letter from the President," *Probe Newsletter* 1, 2 (1 April 1969): 5, PPP; "Schools," *Probe Newsletter* 1, 3 (4 June 1969): 2, PPP; Stanley Zlotkin, "Public Education," *Probe Newsletter* 2, 2 (31 March 1970): 10, PPP; Stanley Zlotkin, interview with author, 19 February 2008, Toronto.
76 Tony Barrett, "Financial Campaign," *Probe Newsletter* 1, 5 (19 September 1969): 4, PPP; "Breadman's Report," *Probe Newsletter* 1, 7 (December 1969): 6, PPP; Tony Barrett, "Membership Fees," *Probe Newsletter* 2, 3 (8 August 1970): 2, PPP.
77 Terry O'Malley, interview with author, 8 July 2008, conducted by telephone.
78 "All's Well That's Financed Well," *Probe Newsletter* 2, 1 (January 1970): 1, PPP.
79 Peter Middleton, interview with author, 21 February 2008, Toronto.
80 Rob Mills, interview with author, 25 September 2008, conducted by telephone.
81 Peter Love, interview with author, 10 March 2008, Toronto.
82 W.B. Harris to Claude Bissell, 7 January 1970, Board of Governors, A77-0019/035, UTA.
83 Rob Mills, letter to supporters, 24 April 1970, Zoology, A77-0019/024, UTA.
84 Rob Mills, letter to supporters, 14 May 1970, Zoology, A77-0019/024, UTA.
85 Peter Love to Claude Bissell, 13 March 1970, Zoology, A77-0019/024, UTA.

86 Rob Mills, letter to supporters, 24 April 1970, Zoology, A77-0019/024, UTA.
87 Brian Kelly, interview with author, 12 January 2009, conducted by telephone.
88 Mark Dowie, *Losing Ground: American Environmentalism at the Close of the Twentieth Century* (Cambridge, MA: The MIT Press, 1995): 106-7.
89 Bosso, *Environment, Inc.*, 36-39, 108; Robert Gottlieb, *Forcing the Spring: The Transformation of the American Environmental Movement* (Washington, DC: Island Press, 2005), 154.
90 Zena Cherry, "FOE Focusses [sic] on Pollution Problems," *Globe and Mail*, 16 April 1970, W2; Kelly, "How to Form Your Own Pollution Probe Group," Projects/Reports/Submissions 1970s, PPP.
91 "1971 Year-End Report," December 1971, PPP; Brian Kelly, "The Environmental Liaison Conference," *Probe Newsletter* 1, 6 (31 October 1969): 3, PPP; "High School Members," *Probe Newsletter* 1, 7 (December 1969): 1, PPP; "Carleton Students' 'Pollution Probe' Goes into Action," Ottawa *Citizen*, 10 December 1969, 38; Robert Paehlke, interview with author, 19 April 2010, conducted by telephone; Eric Hellman, interview with author, 12 January 2010, conducted by telephone.
92 "Summer Project '70: A Proposal from Pollution Probe," 1970, Proposals re Summer 1970, F1058 MU7330, AO.
93 "Summer Project '70 – A Proposal from Pollution Probe," *Probe Newsletter* 2, 2 (31 March 1970): 2, PPP.
94 "Summer Project '70: Final Report," 20 October 1970, 24, PPP; "A Note from the Eco-Financier," *Probe Newsletter* 2, 5 (n.d.): 2, PPP.
95 "Summer Project '70: Final Report," ibid.
96 Samuel P. Hays, "The Mythology of Conservation," in *Perspectives on Conservation*, ed. Henry Jarrett (Baltimore: Johns Hopkins University Press, 1958), 41-42.
97 Paul Ehrlich, *The Population Bomb* (New York: Ballantine Books, 1968), 11.
98 Quoted on last page advertisement in Ehrlich, *The Population Bomb*.
99 Gottlieb, *Forcing the Spring*, 382; Roy Beck and Leon Kolankiewicz, "The Environmental Movement's Retreat from Advocating US Population Stabilization (1970-98): A First Draft of History," in *Environmental Politics and Policy, 1960s-1990s*, ed. Otis L. Graham Jr. (University Park: Pennsylvania State University Press, 2000), 123.
100 Henry Regier and J. Bruce Falls, eds., *Exploding Humanity: The Crisis of Numbers* (Toronto: House of Anansi Press, 1969), 7; "Cadbury, George (1907-95) and Barbara (1910-)," in *Encyclopedia of Birth Control*, ed. Vern L. Bullough (Santa Barbara, CA: ABC-CLIO, 2001), 41.
101 Chris Plowright, interview with author, 30 January 2008, conducted by telephone.
102 Chris Plowright, untitled article, *Probe Newsletter* 2, 2 (31 March 1970): 2, PPP.
103 Donald Worster, *Nature's Economy: A History of Ecological Ideas*, 2nd ed. (Cambridge: Cambridge University Press, 1994), 354.
104 Ralph Brinkhurst, interview with author, 8 August 2008, conducted by telephone.
105 John P. Robarts to Mary Ambrose, 28 May 1970, Zero Population Growth 1970-71, F1058 MU7341, AO.
106 "Zero Population Growth Toronto Program and Budget 1970/71," 12, Zero Population Growth 1970-71, F1058 MU7341, AO.
107 Chris Plowright, interview with author, 30 January 2008, conducted by telephone.
108 "Whither Probe? A Play in IV Parts," 17 February 1971, 3, Whither Probe? F1058 MU7329, AO.
109 Dennis Power to Ryan O'Connor, personal correspondence, 21 July 2010.

110 Chris Plowright, interview with author, 30 January 2008, conducted by telephone.
111 "Plea for Funds Ignored Pollution Group Says," Toronto *Star,* 10 April 1970, 7.
112 "Alderman Plans Leave-Car-at-Home Week," Toronto *Star,* 6 April 1970, 5.
113 "Officials Say Pollution Probe Plan to Close Bay Street Won't Work," *Globe and Mail,* 23 May 1970, 5; "City Committee Votes to Limit Bay Traffic," *Globe and Mail,* 26 May 1970, 5; "Cass Opposes Traffic Tests in Leave Car Home Week," Toronto *Star,* 30 May 1970, 6; "Committee Rejects Closing of Bay Street," *Globe and Mail,* 9 June 1970, 5.
114 "Leave Car Week Is Rescheduled," *Globe and Mail,* 27 June 1970, 5; "Leave the Car at Home Week – Cancelled," 26 June 1970, Pollution – General 1969-, F1058 MU7338, AO.
115 John D. McCarthy and Mayer N. Zald, "Resources Mobilization and Social Movements: A Partial Theory," *American Journal of Sociology* 82, 6 (May 1977): 1212-41.

CHAPTER 3: BUILDING AN ENVIRONMENTAL COMMUNITY

1 Peter Middleton, interview with author, 21 February 2008, Toronto.
2 B. Anthony Barrett, "A Note from the Eco-Financier," *Probe Newsletter* 2, 5 (September 1970): 3, PPP.
3 Ibid.
4 Pollution Probe, "1971 Year-End Report," 14, Annual Reports, PPP.
5 "Co-ordinator's Meeting," 1 June 1971, Minutes of Meetings, F1058 MU7328, AO; untitled press release, 10 February 1971, Financial Statements, F1058 MU7329, AO; Graham Staffen, "Noise Pollution Is Running Amuck: Probe," *The Varsity,* 14 January 1972, 11; Terry Alden, "Radiation Pollution," *Probe Newsletter* 2, 5 (September 1970): 11-14, PPP.
6 Tony Barrett, "Pollution Probe – The Book," *Probe Newsletter* 2, 6 (October 1970): 11, PPP.
7 "100 Hold Candle-Lit Vigil for Ontario Parks," Toronto *Star,* 22 April 1970, 3; "MPPs Hand Out 'Pure' Detergent for Earth Day," Toronto *Star,* 22 April 1970, 69.
8 Cherry, "FOE Focusses [sic] on Pollution Problems," W2.
9 Quoted in "Capsule to Mark 'Man's Folly' Buried in Anti-Pollution Program," *Globe and Mail,* 15 October 1970, 5. See also "Survival Week and Survival Day," n.d., Survival Week Oct. 1970, F1058 MU7342, AO; "Bad Day to Stress Survival," *Globe and Mail,* 8 October 1970, 5; "Pollution of Statistics Angers Meeting," *Globe and Mail,* 9 October 1970, 5; "Carbon Monoxide Study," *Globe and Mail,* 9 October 1970, 5.
10 "Survival Week and Survival Day," ibid.
11 Bob Attfield, "Survival Week," *Probe Newsletter* 2, 6 (October 1970): 11-12, PPP; "Pollution of Statistics Angers Meeting," 5; "Up and Coming Action Seminars," *Probe Newsletter* 2, 6 (October 1970): 16, PPP; "Ottawa Schools to Observe Survival Day," *Probe News* [Pollution Probe at Carleton University] 1, 7 (September 1970): 6, PPP. While Survival Day is most often associated with Pollution Probe, the event was founded and coordinated nationally by a small Hamilton-based staff.
12 Quoted in Livingston T. Merchant and A.D.P. Heeney, "Canada and the United States: Principles for Partnership," *Atlantic Community Quarterly* 3, 3 (Fall 1965): 387; Robert Bothwell, *Alliance and Illusion: Canada and the World, 1945-1984* (Vancouver: UBC Press, 2007), 229.
13 Quoted in James Laxer, *The Energy Poker Game* (Toronto: New Press, 1970), 1.
14 George P. Schultz and Kenneth W. Dam, *Economic Policy beyond the Headlines* (Chicago: University of Chicago Press, 1998), 182-83; Stephen J. Randall, *Foreign Oil Policy since World*

War I: For Profits and Security, 2nd ed. (Montreal and Kingston: McGill-Queen's University Press, 2005), 283.
15 Brian Kelly and Geoff Mains, "Environmentally Yours," *Probe Newsletter* 2, 6 (October 1970): 13, PPP.
16 Ibid.
17 Ibid., 13, 14.
18 Brian Kelly, "Report on Energy and Resources Project: Phase I," December 1970, 1-2, Energy + Resources Project 1970, F1057 MU7347, AO.
19 "Now that We've Nursed the Hungry Monster through Its Gas Pains, What Will We Feed It Next?" Toronto *Telegram*, 9 November 1970, 22.
20 Brian Kelly and Stanley Gershman, "Canada Needs New Policy on Resource Flow to U.S.," Toronto *Star*, 17 November 1970, 7.
21 Mitchell Sharp to Peter Middleton, 16 October 1970, Energy + Resources Project 1970, F1057 MU7347, AO.
22 Kelly, "Report on Energy and Resources Project: Phase I," 4.
23 Terrence Wills, "Pollution Fight, Resources Sale Linked, Davis Tells U.S. Seminar," *Globe and Mail*, 19 November 1970, 10. For more on the phosphate issue, see Chapter 2.
24 Kelly, "Report on Energy and Resources Project: Phase I," 5.
25 The other members of the energy panel were Dr. F.H. Knelman, a professor of physics and engineering at Sir George Williams University; Dr. S.E. Drugge from the University of Alberta's Department of Economics; Dr. William Fuller, chairman of the University of Alberta's Department of Zoology; Dr. R.C. Plowright of the University of Toronto's Department of Zoology; Dr. S.J. Townsend from the Institute for Aerospace Studies at the University of Toronto; Ontario Hydro economist Larratt Higgins; and Dr. L. Trainor from the Department of Physics at the University of Toronto. Brian Kelly to J.J. Greene, Minister of Energy, Mines, and Resources, 21 June 1971, Water Panel Correspondence 1971, F1058 MU7338, AO; Brian Kelly to members of the Water Resources Panel, 23 November 1971, Water Panel Correspondence 1971, F1058 MU7338, AO; Brian Kelly to members of Water Resources Panel, 19 May 1972, Water Panel Correspondence 1971, F1058 MU7338, AO; "History of Pollution Probe," unpublished document, April 1972, History – Pollution Probe, F1058 MU7338, AO.
26 Martin V. Melosi, *Garbage in the Cities: Refuse, Reform, and the Environment* (Pittsburgh: University of Pittsburgh Press, 2005), 221.
27 "Probe's Recycling Program," *Probe Newsletter* 4, 3 (1 May 1972): 2, PPP; "Phone Book Will Help Build Homes," Toronto *Star*, 31 March 1971, 49; Ian McKerracher to Paul Tomlinson, 24 June 1971, Telephone Book Recycling Collection 1971, F1058 MU7336, AO; "3,000 Old Telephone Books Added to Probe's Total," *Globe and Mail*, 12 April 1970, 5; Gregory Bryce, interview with author, 18 March 2008, conducted by telephone.
28 Gregory Bryce, letter to supporters, 19 April 1971, TRAC Recycling Project, F1058 MU7335, AO.
29 Clive Attwater, Gregory Bryce, Sean Casey, and C. Dana Thomas, "Recycling Project Summer 1971: Final Report," 18 October 1971, iii, Recycling Project 1971, F1058 MU7334, AO; Gregory Bryce, interview with author, 18 March 2008, conducted by telephone; Clive Attwater, interview with author, 3 October 2008, conducted by telephone.
30 Quoted in Attwater, Bryce, Casey, and Thomas, "Recycling Project Summer 1971," 97.
31 Gregory Bryce, "Municipal Paper Collection," July 1974, 46-47, Municipal Paper Collection 1974, F1058 MU7336, AO; "City Agrees to Collect Newspapers for Recycling," 19 July

1971, Recycling Project, F1058 MU7334, AO; "Moore Park Newspaper Recycling Project: Further Information for Volunteers," 21 July 1971, Recycling Project, F1058 MU7334, AO; "Toronto to Collect Newspapers for Recycling; Separation Supported by 82 percent," 30 August 1971, Recycling Project, F1058 MU7334, AO.
32 *"Residents of Moore Park, We Need Your Help!"* 21 July 1971, Recycling Project, F1058 MU7334, AO.
33 "Moore Park Newspaper Recycling Project: Further Information for Volunteers," 21 July 1971, Recycling Project, F1058 MU7334, AO.
34 James MacKenzie, "City to Collect Papers," *Globe and Mail,* 20 July 1971, 5.
35 "Toronto to Collect Newspapers for Recycling: Separation Supported by 82 percent," 30 August 1971, Recycling Project, F1058 MU7334, AO.
36 Bryce, "Municipal Paper Collection," 48-50; "City Losing $8 a Ton Recycling Papers," Toronto *Star,* 8 October 1971, 12; Gregory Bryce to H.F. Atyeo, 25 January 1973, F1057 MU7361, AO; Gregory Bryce, "Wednesday Paper Collections Cancelled," 12 December 1972, Recycling Team Probe, F1058 MU7334, AO; H.F. Atyeo to Committee on Public Works, 30 April 1973, F1057 MU7361, AO; Gregory Bryce, untitled brief, 29 January 1973, TRAC January 1973, F1057 MU7361, AO; Gordon T. Batchelor to Toronto Recycling Action Committee, 9 May 1973, F1057 MU7361, AO.
37 "Pollution Probe Presents Recycling Grant to Federal Government," n.d., Working Comm. on Recycling, F1058 MU7335, AO; Tony Barrett to Orville Carr, 21 October 1971, Ottawa – Caravan, F1058 MU7334, AO; Tony Barrett to M. Mortlock and Joe Warwick, Ottawa – Caravan, F1058 MU7334, AO; "Pollution Probe Van Visits Prescott," Prescott *Journal,* 27 October 1971, n.p., Ottawa – Caravan, F1058 MU7334, AO.
38 "Truckload of Garbage to Show Province It Isn't Rubbish," Toronto *Star,* 11 October 1971, 47; "Caravan on Campus," University of Waterloo *Gazette,* n.p., n.d., Ottawa – Caravan, F1058 MU7334, AO; "Pollution Probe Caravan – Further Information," 10 September 1971, Ottawa – Caravan, F1058 MU7334, AO; "Caravan to Pick Up Tons of 'Garbage,'" Burlington *Gazette,* 6 October 1971, n.p., Ottawa – Caravan, F1058 MU7334, AO; "Pollution Probe Van Visits Prescott," Prescott *Journal,* 27 October 1971, n.p., Ottawa – Caravan, F1058 MU7334, AO.
39 "Garbage Hauled Up to Hill to Spur Government Action," Ottawa *Citizen,* 14 October 1971, 1; "Pollution Probe – Ottawa Participates in National Environmental Awareness Week – October 11-October 15, 1971," 4 October 1971, Ottawa – Caravan, F1058 MU7334, AO.
40 Untitled speech, 14 October 1971, Ottawa – Caravan, F1058 MU7334, AO.
41 "Pollution Probe Gives Big Grant to Ottawa," *Globe and Mail,* 15 October 1971, 9.
42 "Resources Recycling Caravan," *Probe Newsletter* 3, 5 (n.d.): 2, PPP.
43 Ibid.; "13½ Ton Junk Gift Presented to Ottawa," Toronto *Star,* 15 October 1971, 12; "Pollution Probe Gives Big Grant to Ottawa," *Globe and Mail,* 15 October 1971, 9.
44 "Pollution Probe Man Launching New Body," *Globe and Mail,* 5 August 1970, 5; "Pollution Groups Pick Chairman from Toronto," Toronto *Star,* 5 August 1970, 30.
45 Peter Middleton, interview with author, 21 February 2008, Toronto.
46 Ibid.; "Summer Reports," *Probe Newsletter* 3, 3 (n.d.): 1, PPP.
47 "Citizens Organized to Protect the Environment," *Probe Newsletter* 2, 5 (September 1970): 7, PPP; Rob Mills, "The C.O.P.E. Project," *Probe Newsletter* 2, 7 (December 1970): 27-28, PPP; Barry Came, "Thousands of Students Join COPE Information Drive," *Globe and Mail,* 23 November 1970, 5; Peter Middleton, "A Proposal," 28 May 1971, Whither Probe?

F1058 MU7329, AO; "Co-ordinator's Meeting," 1 June 1971, Minutes of Meetings, F1058 MU7328, AO.

48 Alan Levy, "Readers Digest of CELA's History," *Intervenor* 26, 1 (January-March 2001), http://www.cela.ca/article/readers-digest-celas-history.

49 CELRF's founding board of directors demonstrates the foundation's close connection with Pollution Probe, as both Donald Chant and Peter Middleton were included. The other members were Barry Stuart, an Osgoode Hall law professor; Clayton Hudson, a tax lawyer at Shibley Righton LLP; and Eddie Goodman, a prominent Tory insider and founding partner at Goodman and Goodman, now Goodman LLP. Untitled document, 18 July 1971, CELA and CELRF incorporation, 4.11, John Swaigen fonds, Wilfrid Laurier University Archives.

50 Thomas R. Dunlap, *DDT: Scientists, Citizens, and Public Policy* (Princeton, NJ: Princeton University Press, 1981), 143.

51 D. Paul Emond, "'Are We There Yet?' Reflections on the Success of the Environmental Law Movement in Ontario," *Osgoode Hall Law Journal* 46, 2 (Summer 2008): 223.

52 Alan Levy, interview with author, 19 March 2008, Toronto; Levy, "Readers Digest of CELA's History"; "Announcing ... the New, Improved Pollution Complaint Service," *Probe Newsletter* 2, 7 (December 1970): 3-4, PPP.

53 David Estrin, interview with author, 19 March 2008, Toronto; "General Counsel: David Estrin," *Environmental Law News* 1, 1 (1 February 1972): 6.

54 Alan Levy, interview with author, 19 March 2008, Toronto; Levy, "Readers Digest of CELA's History."

55 John Scott, "The Toll of Steady Quarrying on the Sandbanks," *Globe and Mail*, 1 September 1971, 25.

56 Norris Whitney, "Sandbanks Provincial Park," *Globe and Mail*, 7 August 1971, 7.

57 Norris Whitney, "Sandbanks Provincial Park," *Globe and Mail*, 27 July 1971, 6; Scott, "The Toll of Steady Quarrying on the Sandbanks," 25.

58 "A Good Start, but," *Globe and Mail*, 2 December 1971, 6.

59 These are figures provided by Lake Ontario Cement. Scott, "The Toll of Steady Quarrying on the Sandbanks," 25.

60 "Ontario Government Sued for Breach of Trust over Sandbanks Provincial Park," *Environmental Law News* 1, 2 (10 May 1972): 7.

61 Peter Whelan, "Breach of Trust Suit Is Filed against Ontario on Sandbanks," *Globe and Mail*, 9 August 1972, 5.

62 "Campers Picket Firm at Sandbanks," *Globe and Mail*, 6 July 1972, 41.

63 Peter Whelan, "Woman Prosecutes Cement Firm and Sandbanks Hauler," *Globe and Mail*, 20 July 1972, 10.

64 John Slinger, "Sandbanks Report Rejects Dune Peril, Urges 3-Year Deal," *Globe and Mail*, 13 July 1972, 1; Ron Alexander and Larry Green, "A Future for the Sandbanks: A Report on the Sand Dunes of Prince Edward County," 18 September 1972, 5-8, 16-17, 25, Sandbanks 1972-74, F1058 MU7341, AO.

65 "Sandbanks Provincial Park: All Is Quiet – Temporarily," *Environmental Law News* 1, 4 (November 1972): 11-12; "Ontario Supreme Court Rules that Provincial Parks Act Lacks Trust Provisions," *Environmental Law News* 2, 1 (February 1973): 4-7.

66 "Ontario Supreme Court Rules that Provincial Parks Act Lacks Trust Provisions," 7; Vianney Carriere, "Province Will Expropriate Lease on Quarry near Sandbanks Park," *Globe and*

    Mail, 22 March 1973, 1; "Province Expropriates Sandbanks Lease," *Environmental Law News* 2, 2 (April 1973): 21-22.
67 David Estrin, interview with author, 19 March 2008, Toronto. Manning would gain national prominence in the 1980s as defence counsel to abortionist Henry Morgentaler.
68 "Environment Provincial Matter, Court Holds," *Globe and Mail*, 17 March 1973, 5.
69 Peter Middleton, interview with author, 21 February 2008, Toronto; Monte Hummel, interview with author, 23 January 2008, Toronto; Brian Gifford [Ecology Action Centre] to Peter Middleton, 20 June 1973, Ecology Action Centre 1973, F1058 MU7342, AO; Brian Gifford to Tony Barrett, 25 August 1973, Ecology Action Centre 1973, F1058 MU7342, AO; Dale Berry [SPEC] to Peter Middleton, 16 July 1973, Vancouver SPEC 1973, F1058 MU7334, AO; Tony Barrett to Dale Berry, 24 August 1973, Vancouver SPEC 1973, F1058 MU7334, AO; Dale Berry to Tony Barrett, 25 September 1973, Vancouver SPEC 1973, F1058 MU7334, AO.
70 Joyce Rothschild and J. Allen Whitt, *The Cooperative Workplace: Potentials and Dilemmas of Organisational Democracy and Participation* (New York: Cambridge University Press, 1986).
71 This frustration was discussed in a variety of interviews, including: Peter Middleton, interview with author, 21 February 2008, Toronto; Ann Love, interview with author, 4 April 2008, conducted by telephone; Brian Kelly, interview with author, 12 January 2009, conducted by telephone; Monte Hummel, interview with author, 23 January 2008, Toronto.
72 Hummel, ibid.
73 Paul Tomlinson, "The Second Epistol [sic]," 25 May 1971, Whither Probe? F1058 MU7329, AO.
74 Peter Middleton, "A Proposal," 28 May 1971, Whither Probe? F1058 MU7329, AO.
75 Peter Middleton, interview with author, 21 February 2008, Toronto.
76 Monte Hummel, "Robarts and Bottles – The Non-Returnables," *Probe Newsletter* 2, 4 (June 1970): 3, PPP.
77 "The Progressive Conservative Leadership Candidates and the Environment: An Evaluation of Their Views by Pollution Probe," 4 February 1971, Press Releases 1971, PPP.
78 Davis was a great admirer of Chant and later appointed him head of the Ontario Waste Management Corporation. Bill Davis, interview with author, 5 June 2008, conducted by telephone.
79 Peter Middleton, interview with author, 21 February 2008, Toronto.
80 "Davis Offers More Aid to Pollution Probe," *Toronto Star*, 4 August 1971, 2.
81 As historian Frank Zelko notes, Pollution Probe was among the groups that provided financial support to the Don't Make a Wave Committee's campaign. Zelko, *Make It a Green Peace!*, 71.
82 Rex Weyler, *Greenpeace: How a Group of Ecologists, Journalists and Visionaries Changed the World* (Vancouver: Raincoast Books, 2004), 129-30; *The Varsity*, 5 November 1971, 8-9.
83 "Wrapped Pig's Head the Final Blast Protest," *Globe and Mail*, 6 November 1971, 5; "Amchitka," *Probe Newsletter* 3, 5 (n.d.): 7, PPP.
84 Ann Love, interview with author, 4 April 2008, conducted by telephone.
85 Ibid.; "Wrapped Pig's Head the Final Blast Protest," 5.
86 Ann Love, interview with author, 4 April 2008, conducted by telephone.
87 A. Paul Pross, "Pressure Groups: Adaptive Instruments of Political Communication," in *Pressure Group Behaviour in Canadian Politics*, ed. A. Paul Pross (Toronto: McGraw-Hill Ryerson, 1975), 11-15.

Chapter 4: Probe's Peak

1 Frank Summerhayes, "Pollution Probe," *Globe and Mail,* 30 June 1972, 6.
2 Untitled press release, 6 June 1972, Urban Team: Probe – Rules of the Game 1972, F1058 MU7340, AO.
3 *Rules of the Game: A Handbook for Tenants and Homeowners* (Toronto: Pollution Probe Foundation, 1973), Urban Project – Misc. Materials, F1058 MU7340, AO.
4 James McKenzie, "Pollution Probe Picks on Property Politics," *Globe and Mail,* 16 June 1972, 41.
5 "Civic Meetings of Interest to Residents of Metro," *Globe and Mail,* 9 November 1970, 5; Monte Hummel, "Education," *Probe Newsletter* 2, 7 (December 1970): 26, PPP; Ruth Kelly and Larry Erickson, "The Education Team," n.d., 1-2, Education Program 1974, F1058 MU7343, AO; "Whether or Not to Hire a Person for the Education Project," n.d., Education Program 1974, F1058 MU7348, AO.
6 Task Force Hydro brochure, quoted in "Task Force Hydro Brief," *Probe Newsletter* 4, 2 (n.d.): 2, PPP.
7 "Brief to Task Force Hydro," 11 February 1972, 4-19, Ont. Hydro 1969-72, F1058 MU7338, AO.
8 "Advisory Committee on Energy," Toronto *Star,* 9 September 1971, 42.
9 Pollution Probe, "Brief to Ontario Advisory Committee on Energy," July 1972, iii-iv, Energy + Resources Project 1972-, F1057 MU7348, AO.
10 Pollution Probe, *"It's Time to Think of the Future,"* 18 July 1972, Can. Energy + Resources Policy 1970-, F1057 MU7349, AO.
11 Task Force Hydro, *Hydro in Ontario: A Future Role and Place* (Toronto: Ontario Legislative Assembly Committee on Government Productivity, 1973).
12 Task Force Hydro, *Nuclear Power in Ontario* (Toronto: Ontario Legislative Assembly Committee on Government Productivity, 1973).
13 Task Force Hydro, *Hydro in Ontario: An Approach to Organization* (Toronto: Ontario Legislative Assembly Committee on Government Productivity, 1972); Task Force Hydro, *Hydro in Ontario: Financial Policy and Rates* (Toronto: Ontario Legislative Assembly Committee on Government Productivity, 1973).
14 Advisory Committee on Energy, *Energy in Ontario: The Outlook and Policy Implications,* Volume 1 (Toronto: Ontario Legislative Assembly Committee on Government Productivity, 1973); David Crane, "Ontario Facing Energy Shortage," Toronto *Star,* 15 January 1973, 1, 4. Volume 2 of the ACE report was released on 5 March 1973.
15 "A Problem but No Policy," Toronto *Star,* 25 January 1973, 6.
16 "Letter from Henry A. Regier," 19 December 1972, Appendix F, in Advisory Committee on Energy, *Energy in Ontario,* 38.
17 For more on the Mackenzie Valley Pipeline and the surrounding debate, see Robert Page, *Northern Development: The Canadian Dilemma* (Toronto: McClelland and Stewart, 1986), 24-58, 97-98, 104-5; Thomas Berger, *Northern Frontier, Northern Homeland: The Report of the Mackenzie Valley Pipeline Inquiry* (Toronto: Douglas and McIntyre, 1988).
18 Page, then a history professor at Trent University, chaired the Committee for an Independent Canada's anti-pipeline efforts, and thus had a first-hand view of Pollution Probe's work. Page, ibid., 37.
19 "Pollution Probe's Arctic Campaign," 28 March 1972, Arctic Campaign – Northern Gas Pipeline 1972, F1057 MU7349, AO.

20 "Background Statement on the Arctic," 28 March 1972, Energy and Resources Project, F1057 MU7348, AO.
21 "Proposal and Recommendations," 27 June 1983, Mackenzie Valley Gas Pipeline 1973, PPP.
22 The Energy and Resources Team's expenditures amounted to $38,214, out of a total expenditure of $157,997. These numbers do not include the cost of the Caravan Team, which was designated a "special project of Pollution Probe" and thereby audited separately. Its costs, which were underwritten by project sponsors, were $150,358, which would increase Pollution Probe's total expenditures for the fiscal year to $308,355. "Probe Annual Report 1973," 5, Annual Reports, PPP.
23 The Task Force was initially chaired by A.S. Bray, the former senior assistant deputy minister and special consultant to the minister of industry and tourism. Bray's untimely death led to the naming of his replacement, R.H. Woolvett of Brewers Warehousing Company Ltd., on 3 October 1973. Others named to the Task Force were K.M. Bethune of the Metal Container Manufacturers' Advisory Council, Mrs. W.A. Brechin of the Consumers' Association of Canada, H.E. Dalton of the Glass Container Council of Canada, L.G. Jamison of the Packaging Association of Canada, Mrs. R.L. MacMillan of the Conservation Council of Ontario, Ian McKerracher of the Municipal Engineers' Association, H.D. Paavila of the Canadian Pulp and Paper Association, E.G. Salmond of the Society of the Plastics Industry of Canada, and F.M. Woods of the Association of Municipalities of Ontario. *General Report of the Solid Waste Task Force to the Minister of the Environment, Part 1* (Toronto: Ministry of the Environment, 1974), 13-14, 21.
24 Ibid., 23.
25 "The Provincial Solid Waste Task Force," *Probe Newsletter* 4, 6 (November 1972): 2, PPP.
26 *General Report of the Solid Waste Task Force to the Minister of the Environment, Part 1*, 21; Peter Love, interview with author, 10 March 2008, Toronto.
27 Peter Love, "Preliminary Submission of Pollution Probe to the Ontario Task Force on Solid Waste," November 1972, 4-7, Recycling Team: Probe, F1058 MU7334, AO.
28 Peter Love, interview with author, 10 March 2008, Toronto.
29 Gregory Bryce to the Toronto Recycling Action Committee, 10 January 1973, F1057 MU7361, AO; Gregory Bryce, interview with author, 18 March 2008, conducted by telephone; Peter Love, interview with author, 10 March 2008, Toronto.
30 *General Report of the Solid Waste Task Force to the Minister of the Environment, Part 1*, 19-20.
31 "Probe Comments on the Report of the Solid Waste Task Force," 19 December 1974, Beverage Packaging Working Group – Milk Packaging 1973, F1058 MU7336, AO.
32 Quoted in Diane Humphries, *We Recycle: The Creators of the Blue Box Programme* (Toronto: Pollution Probe, 1997), 6, http://www.pollutionprobe.org/Reports/we%20recycle.pdf.
33 Robert Duffy, "The Throwaway Habit Isn't Easy to Kick," Toronto *Star*, 25 March 1976, D1.
34 Gregory Bryce, "A First Step," *Globe and Mail*, 14 March 1973, 43; Gregory Bryce, interview with author, 18 March 2008, conducted by telephone; Peter Love, interview with author, 10 March 2008, Toronto; Peter Middleton, interview with author, 21 February 2008, Toronto.
35 The remaining component of the solution, of course, was a commitment to reducing throughput. As Bryce noted: "Technological solutions ... are only part of the answer. In Franklin we had occasion to reflect on the marketing practices and the values that cause garbage to increase much faster than population. It was depressing at the plant to see a large nearly-new toy truck being dumped at the receiving ramp." Bryce, "A First Step."

36 James Auld to Gordon Burbidge [Pollution Probe], 25 May 1972, F1057 MU7361, AO.
37 Peter Whelan, "$17 Million Program to Recycle Garbage Planned in Ontario," *Globe and Mail*, 25 October 1974, 1; "Control of Litter-Producing Items Poses Political Problem in Ontario," *Globe and Mail*, 25 October 1974, 31; "Garbage Plan," *Globe and Mail*, 25 October 1974, B1; "A Long-Range Plan to Reclaim Our Waste," *Globe and Mail*, 29 October 1975, 6.
38 "Garbage Team Activities and Expenses: October 1975 – September 1976," 15, Garbage Team Report 1975-76, F1058 MU7336, AO.
39 "Lynn Spink" is the name used when citing interviews conducted for this book, and "Marilyn Cox" is used in reference to her historical activities working at Pollution Probe.
40 Marilyn Cox to Ross Perry, 23 June 1972, Community Planning Assoc. 1972-73, F1058 MU7340, AO.
41 James Lemon, *Toronto since 1918: An Illustrated History* (Toronto: James Lorimer, 1985), 162.
42 "The Urban Team: Our Position," 5 April 1972, Urban Project F1058 MU7340, AO.
43 Lynn Spink, interview with author, 12 March 2008, Toronto.
44 Ibid.
45 Ibid.
46 Carl Dreisziger and Tom Murphy, "A Pollution Probe Brief on the Environmental Implications of the Proposed Spadina Rapid Transit System," 15 March 1972, Urban Team: Probe – Rules of the Game 1972, F1058 MU7340, AO.
47 Tom Murphy, "A Brief to the Ontario Municipal Board Respecting the Selection of a Subway Route in the Spadina Corridor," 18 April 1973, Transportation Proposal Wheeling and Dealing 74, F1058 MU7341, AO.
48 Lynn Spink, interview with author, 12 March 2008, Toronto; "Industry in Residential Areas Report," October 1973, January-February 1974, Urban Project – Progress Reports 1973, F1058 MU7341, AO.
49 *Probe Bulletin*, 1 October 1974, 1, PPP.
50 Lynn Spink, interview with author, 12 March 2008, Toronto.
51 Gregory Bryce, interview with author, 18 March 2008, conducted by telephone.
52 Peter Middleton, interview with author, 21 February 2008, Toronto.
53 For more on the history of environmental justice, see Gottlieb, *Forcing the Spring*, 307-46; Robert D. Bullard and Beverly H. Wright, "The Quest for Environmental Equity: Mobilizing the African-American Community for Social Change," in *American Environmentalism: The U.S. Environmental Movement, 1970-1990*, ed. Riley E. Dunlap and Angela G. Mertig (Philadelphia: Taylor and Francis, 1992), 39-49.
54 A. Paul Pross, "Canadian Pressure Groups," in *Pressure Groups*, ed. Jeremy J. Richardson (New York: Oxford University Press, 1993), 151-52; Peter Middleton, interview with author, 21 February 2008, Toronto; Ann Rounthwaite, interview with author, 10 July 2008, conducted by telephone.
55 "Probe Special Report – No. 1: Canadian Gypsum Company," March 1972, 14, Probe – Canadian Gypsum 1972, F1058 MU7334, AO.
56 Alden Baker, "York Wants Action on Gypsum," *Globe and Mail*, 14 March 1972, 5; Peter Whelan, "Cleanup Schedule for Gypsum Plant Demanded by York Health Board," *Globe and Mail*, 22 March 1972, 31.
57 "Gypsum Pollution Evidence Lacking, Auld Says," *Globe and Mail*, 23 March 1972, 5.
58 This pledge resulted in two temporary production shutdowns during the week of 17 April 1972. "Firm Holds to Pledge, Closes for Pollution," *Globe and Mail*, 22 April 1972, 2.

59 "Pollution Probe Guide to Pollution Fighting: Any Group Can!" April 1972, 18, Probe – Canadian Gypsum 1972, F1058 MU7339, AO.
60 "Other Probe News," *Probe Newsletter* 4, 3 (1 May 1972): 7, PPP.
61 Peter Middleton to Ryan O'Connor, personal correspondence, 19 July 2010.
62 Untitled document, September 1971, Minutes of Meetings, F1058 MU7328, AO; "Proposal that Probe Should Become Involved in Regional and Land Use Planning," n.d., Land Use – Project TOE 73, F1058 MU7340, AO; David Wood, interview with author, 20 March 2008, conducted by telephone; Frances Frisken, *The Public Metropolis: The Political Dynamics of Urban Expansion in the Toronto Region, 1924-2003* (Toronto: Canadian Scholars' Press, 2007), 127-30; Julie-Ann Coudreau, Roger Keil, and Douglas Young, *Changing Toronto: Governing Urban Neoliberalism* (Toronto: University of Toronto Press, 2009), 115.
63 "Brief to the Pickering Airport Inquiry Commission," 8 April 1974, Pickering Airport 1974, F1058 MU7340, AO.
64 *The Tail of the Elephant: A Guide to Regional Planning and Development in Southern Ontario* (Toronto: Pollution Probe Foundation, 1974), 6, "The Tail of the Elephant" Regional Planning 1974, F1058 MU7341, AO.
65 "The Short Happy Life of the Land Use Project," *Probe Bulletin*, 1 October 1974, 2, Probe Bulletins, PPP.
66 Of that amount, $84,000 came from the provincial Ministry of the Environment, $50,000 from the Richard Ivey Foundation, $30,000 from Canada Packers Ltd., and $25,000 each from Carling O'Keefe and the Imperial Tobacco-funded White Owl Conservation Awards. "Project 73 Final Report and Probe 74 Proposal," 23 August 1973, 1, Project 73, PPP; "Project 73," n.d., Project 73, PPP; "Pollution Probe at the University of Toronto Annual Report 1973 and Five Year Review," 1, Annual Reports, PPP; Untitled press release, n.d., Press Releases 1973, PPP.
67 "Pollution Probe Annual Report 1973," 1, Annual Reports, PPP.
68 Joe Warwick, interview with author, 6 August 2008, conducted by telephone.
69 "The Pollution Probe Foundation Financial Statements: September 30, 1973," 4-5, Financial Statements, PPP; "The Pollution Probe Foundation Financial Statements: September 30, 1974," 5-6, Financial Statements, PPP; "The Pollution Probe Foundation Financial Statements: September 30, 1975," 4-5, Financial Statements, PPP; "Whither Probe," 7 January 1974, Whither Probe? '74, F1058 MU7329, AO; "Report on Kilcoo 1974," May 1974, Organizational Development, F1058 MU7329, AO.
70 Untitled article, *Probe Bulletin*, 1 October 1974, 1, PPP.
71 "Peter Middleton: The Entrepreneur of Renewables," *Saturday Night*, March 1977, 32-33; Alan MacEachern, *The Institute of Man and Resources: An Environmental Fable* (Charlottetown: Island Studies Press, 2003), 31-33, 38, 120; "Ian Efford/Brian Kelly: The Panjandrums of Zero Growth," *Saturday Night*, March 1977, 34-35.
72 Peter Love, interview with author, 10 March 2008, Toronto.

CHAPTER 5: THE CHANGING ENVIRONMENTAL LANDSCAPE

1 Paehlke, "Eco-History," 18.
2 Ross Howard, "Pollution Probe's Alive and Fighting for Environment," Toronto *Star*, 19 May 1977, F7.

3 Michael Kanter, "Group Continues Pollution Probe," *The Varsity*, 28 September 1977, 12. A review of the *Globe and Mail* reveals that Pollution Probe was mentioned in eighty-four articles in 1970 and sixty-four in 1972; it appeared three times in 1977, and twelve times in 1979.
4 Brian Kelly to Phil Lind, 14 December 1973, Energy Probe – Establishment Energy + Resources Team, F1057 MU7347, AO.
5 Sanford Osler, interview with author, 12 August 2008, conducted by telephone.
6 Untitled document, 21 January 1974, Energy Probe – Establishment Energy + Resources Team, F1057 MU7347, AO.
7 "Statement by Dr. Donald Chant," 16 January 1975, Energy Probe – Establishment Energy + Resources Team, F1057 MU7347, AO.
8 Sanford Osler, interview with author, 12 August 2008, conducted by telephone.
9 Ibid.
10 Michael D. Mehta, *Risky Business: Nuclear Power and Public Protest in Canada* (Toronto: Lexington Books, 2005), 38-39; Michael Maurice Dufrense, "'Let's Not Be Cremated Equal': The Combined Universities Campaign for Nuclear Disarmament, 1959-1967," in Palaeologu, *The Sixties in Canada*, 9-64.
11 Barry Spinner, interview with author, 29 July 2009, conducted by telephone.
12 Neil B. Freeman, *The Politics of Power: Ontario Hydro and Its Government, 1906-1995* (Toronto: University of Toronto Press, 1996), 163; Robert Macdonald, "Energy, Ecology, and Politics," in *Ecology versus Politics in Canada*, ed. William Leiss (Toronto: University of Toronto Press, 1979), 196; Jamie Swift and Keith Stewart, *Hydro: The Decline and Fall of Ontario's Electric Empire* (Toronto: Between the Lines, 2004), 20.
13 Quoted in Walt Patterson, "Porter with Nuclear Reservations," *New Scientist*, 2 November 1978, 362.
14 "Energy Probe Program: October 1976 – September 1977," n.d., 4, Energy Probe – Establishment Energy + Resources Team, F1057 MU7347, AO; Bill Peden, "Submission to the Royal Commission on Electric Power Planning," 14 November 1975, Porter Commission – Ont. Electric Planning 1975-, F1057 MU7352, AO.
15 *CANDU: An Analysis of the Canadian Nuclear Program* (Toronto: Pollution Probe Foundation, 1975), v, Candu – Technical Handbook 1975-76, F1057 MU7356, AO.
16 "Energy Probe Program: October 1976 – September 1977," n.d., 14-15, Energy Probe – Establishment Energy + Resources Team, F1057 MU7347, AO; "Ecology House Opens Its Doors," *Probe Post* 3, 3 (September-December 1980): 1; "The Pollution Probe Foundation Financial Statements: September 30, 1974," 5-6, Financial Statements, PPP; "The Pollution Probe Foundation Financial Statements: September 30, 1975," 4-5, Financial Statements, PPP; "The Pollution Probe Foundation Financial Statements: September 30, 1976," 4, Financial Statements, PPP.
17 "People," *Probe Bulletin* 7 (May 1975): 2, PPP; "Director Leaves," *Probe Bulletin* 11 (May 1976): 2, PPP.
18 David Wood, interview with author, 20 March 2008, conducted by telephone; "Probe on Pickering: 'Take a Train!'" *Probe Bulletin* 8 (August 1975): 1, PPP; "Pickering Airport," *Probe Bulletin* 9 (November 1975): 2, PPP; Frisken, *The Public Metropolis*, 159.
19 Eric Hellman, interview with author, 12 January 2010, conducted by telephone; Stef Donev, "Metro Garbage Can't Go to Hope," Toronto *Star*, 16 May 1975, A1; Michael Best, "Metro's Garbage: Bad News and Good," Toronto *Star*, 21 May 1975, D1; Brian Dexter, "Pickering Garbage Site Approved by Province," Toronto *Star*, 17 February 1976, B2.

20 "Pollution Probe's Action Campaign 1977," n.d., 3, Annual Report 1977, PPP; "Pop Posse Kits to Help You Report Soft Drink Sellers," Toronto *Star*, 3 May 1977, C2; "Brite! Brite! Brite! Pollution Probe's Boomerang Information Sheet," n.d., Factsheets/Brochures 1977, PPP; "Probe Gives Pat on the Back for Environmental Effort," 20 October 1977, Press Releases 1977, PPP; "Probe Pounces on Corporate Culprits," 18 January 1978, Press Releases 1978, PPP; "Pop Posse Pays Off," *Environmental Newsletter*, 18 January 1978, 2, Factsheets/Brochures 1978, PPP.
21 Robert Gibson, interview with author, 16 October 2009, conducted by telephone.
22 "Energy Probe Program: October 1976-September 1977," n.d., 3, 14, Energy Probe – Establishment Energy + Resources Team, F1057 MU7347, AO; "Ecology House Opens Its Doors," *Probe Post* 3, 3 (September-December 1980): 1; "Federal Government Provides Assistance for Toronto 'Ecology House' Project," 3 May 1978, Press Releases 1978, PPP.
23 *Ecology House Newsletter*, 1 May 1980, Ecology House 1981, PPP; Linda Stone, "Old House Points Way to Energy-Saving Future," *Globe and Mail*, 28 November 1980, BL4; Ann Finlayson, "Open the House, Close the Door," *Maclean's*, 24 November 1980, n.p.
24 *Canada as a Conserver Society: An Agenda for Action* (Ottawa: Science Council of Canada, 1978), 9.
25 John B. Robinson and D. Scott Slocombe, "Exploring a Sustainable Future for Canada," in *Life in 2030: Exploring a Sustainable Future for Canada*, ed. John B. Robinson (Vancouver: UBC Press, 1996), 5.
26 Lawrence Solomon, interview with author, 14 August 2008, conducted by telephone.
27 Lawrence Solomon, *The Conserver Solution* (Toronto: Doubleday Canada, 1978), 4.
28 "What They're Saying about *The Conserver Solution*" [promotional brochure], n.d., Larry Solomon, F1057 MU7347, AO.
29 Chris Conway, interview with author, 14 July 2009, conducted by telephone.
30 Solomon, *The Conserver Solution*, 184.
31 Chris Conway to Ryan O'Connor, personal correspondence, 9 September 2010.
32 David Brooks, interview with author, 13 July 2009, conducted by telephone.
33 Italics in original. David Brooks, *Zero Energy Growth for Canada* (Toronto: McClelland and Stewart, 1981), viii, x.
34 Weyler, *Greenpeace*, 252.
35 John Bennett, interview with author, 8 July 2008, conducted by telephone.
36 Ibid.; Dan McDermott, interview with author, 26 June 2008, conducted by telephone.
37 McDermott, ibid.
38 John Bennett, interview with author, 8 July 2008, conducted by telephone.
39 Ibid.
40 Douglas Saunders, interview with author, 9 August 2008, conducted by telephone.
41 Tony McQuail, interview with author, 19 August 2008, conducted by telephone.
42 Douglas Saunders, interview with author, 9 August 2008, conducted by telephone.
43 "Greenpeace Sneaks into N-Plant," *Globe and Mail*, 12 July 1977, 4; Weyler, *Greenpeace*, 510-11; John Bennett, interview with author, 8 July 2008, conducted by telephone; Saunders, ibid.; Tony McQuail, interview with author, 19 August 2008, conducted by telephone.
44 "Greenpeace Sneaks into N-Plant," 4.
45 "Ontario Okays Nuclear Plant near Oshawa," Toronto *Star*, 19 July 1977, A2; Freeman, *Politics of Power*, 164; Arthur Johnson, "Impact Testing Issue Seized by Opposition," *Globe and Mail*, 24 August 1977, 5.
46 Quoted in "Ontario Okays Nuclear Plant near Oshawa," A2.

47 "Greenpeace Protests at Darlington Site," CBC Digital Archives, 1 October 1977, http://archives.cbc.ca/science_technology/energy_production/clips/915/.
48 James Jefferson, "12 Arrested in Anti-Nuclear March," *Globe and Mail*, 3 October 1977, 9; Weyler, *Greenpeace*, 511.
49 Dan McDermott, interview with author, 26 June 2008, conducted by telephone.
50 Douglas Saunders, interview with author, 9 August 2008, conducted by telephone.
51 "Probe Stages Environmental Assessment Hearing on Darlington Nuclear Station," *Probe Environment Newsletter* 17 (November 1977): 5, Projects/Reports/Submissions 1977, PPP.
52 Jan Marmorek went by this name while working at Energy Probe but reverted to her maiden name (Jan McQuay) after leaving Energy Probe. Others funding the project included Halifax's Ecology Action Centre, Pollution Probe Ottawa, the Prairie Defence League, the Public Interest Coalition for Energy Planning, STOP Edmonton, and STOP Montreal. Jan Marmorek, *Everything You Wanted to Know about Nuclear Power (but Were Afraid to Find Out!)* (Toronto: Pollution Probe Foundation, 1978), 2, 6, 8-9, Projects/Reports/Submissions 1978, PPP.
53 Freeman, *Politics of Power*, 164-65.
54 "Energy Probe Responds to Commission's Nuclear Report," 28 September 1978, Porter Commission – Ont. Electric Planning 1975-, F1057 MU7352, AO.
55 Gottlieb, *Forcing the Spring*, 239-40.
56 Lawrence Solomon, interview with author, 14 August 2008, conducted by telephone. The actual title of the newsletter was "It's no longer a movie: it just happened in Pennsylvania (and it could happen here)," n.d., Energy Probe Newsletters, F1057 MU7347, AO.
57 These numbers would hold steady through 1983. Ronald Babin, *The Nuclear Power Game* (Montreal: Black Rose Books, 1985), 219.
58 John Munch, "Nuclear Protesters in Tower Sit-in," Toronto *Star*, 1 June 1979, A1.
59 Ross Howard and John Munch, "66 Arrests at Peaceful Anti-Nuke Protest," Toronto *Star*, 3 June 1979, A1, A4.
60 Quoted in Joyce Rowlands, "Nuclear Protest In Public 'Interest,'" Toronto *Star*, 6 June 1979, C8.
61 Freeman, *Politics of Power*, 165-66.
62 "The Pollution Probe Foundation Financial Statements September 30, 1978," Financial Statements, PPP; "The Pollution Probe at the University of Toronto Financial Statements September 30, 1979," Financial Statements, PPP; "How Probe Works," *Probe Bulletin* 10 (February 1976): 2, PPP; Marilyn Aarons, "Energy Probe's Board Update #2," 16 May 1980, 2, Board of Directors, Advisors, F1057 MU7347, AO; "Low income measures, low income after tax cut-offs and low income after-tax measures," Statistics Canada, 7 April 2004, http://www.statcan.gc.ca/pub/13f0019x/13f0019x1997000-eng.pdf.
63 David Coon, interview with author, 18 August 2008, conducted by telephone.
64 Marilyn Aarons, "Board Update #3," 12 August 1980, 1, Board of Directors, Advisors, F1057 MU7347, AO.
65 Lawrence Solomon, interview with author, 14 August 2008, conducted by telephone.
66 Jan McQuay, interview with author, 23 October 2009, conducted by telephone; Marilyn Aarons, interview with author, 29 November 2009, conducted by telephone; David Coon, interview with author, 18 August 2008, conducted by telephone; Babin, *The Nuclear Power Game*, 219.
67 Coon, ibid.
68 Jan McQuay, interview with author, 23 October 2009, conducted by telephone; Lawrence Solomon, interview with author, 14 August 2008, conducted by telephone.

69 Marilyn Aarons, "Energy Probe's Board Update #2," 16 May 1980, 2, Board of Directors, Advisors, F1057 MU7347, AO.
70 Marilyn Aarons, "Board Update #3," 12 August 1980, 1, Board of Directors, Advisors, F1057 MU7347, AO. The Board of Advisors was renamed the Board of Directors sometime before autumn 1976. "Pollution Probe Foundation Annual Report 1976-76," Annual Reports, PPP.
71 Janet Wright, "Pollution Probe Foundation Minutes of Meetings," 24 September 1980, 3, Pollution Probe/Energy Probe Split 1980, PPP.
72 Ibid.
73 Lawrence Solomon, interview with author, 14 August 2008, conducted by telephone. See also B.D. Katchen [lawyer at Shibley, Righton and McCutcheon] to Norm Rubin, 30 October 1980, E.P. Research Foundation 1980-, F1057 MU7347, AO; "Pollution Probe Foundation Annual Report 1980-1981," 1, Annual Reports, PPP.
74 David Coon, interview with author, 18 August 2008, conducted by telephone.
75 "The Pollution Probe Foundation Financial Statements: September 30, 1980," 4, Financial Statements, PPP; Colin Isaacs, interview with author, 18 August 2008, conducted by telephone.
76 David Brooks, interview with author, 13 July 2009, conducted by telephone.
77 Lawrence Solomon, interview with author, 14 August 2008, conducted by telephone.
78 This quote from Energy Probe's 1983 annual report was found in Elaine Dewar, *Cloak of Green* (Toronto: James Lorimer, 1995), 361.
79 Energy Probe considers hydro to be too harmful to the environment to endorse. Instead, it supports oil, gas, and coal as energy sources. Ibid., 358. Solomon has defended his organization from such criticism, saying that "Energy Probe has received no donations from the fossil fuel industry since 1982." According to the Energy Probe Research Foundation website, the majority of its funding comes from individual donors, foundations, the sale of publications, and fees charged for speaking engagements and expert testimony at hearings. Regardless of the source of its funding, Energy Probe and its sister organizations' focus on property rights and free market economics has made it a veritable pariah within the environmental movement. Lawrence Solomon, "Letter: Foreign Donations OK: Suzuki Foundation," *Financial Post*, 19 December 2013, http://opinion.financialpost.com/2013/12/19/letter-foreign-donations-ok-suzuki-foundation/; "Our Funding," Energy Probe Research Foundation, http://epresearchfoundation.wordpress.com/our-funding/.
80 According to Brooks, Solomon demanded the employee be fired because he did not value her educational background in sociology. Further demonstrating the close-knit nature of the first-wave environmental community, Brooks would next join the growing roster at Peter Middleton and Associates. David Brooks, interview with author, 13 July 2009, conducted by telephone.
81 Chris Conway, interview with author, 14 July 2009, conducted by telephone.
82 "Probe's Last Word on 'Recycling,'" August 1978, Projects/Reports/Submissions 1978, PPP.
83 Peter Whelan, "$17 Million Program to Recycle Garbage Planned in Ontario," *Globe and Mail*, 25 October 1974, 1.
84 Harold Crooks, *Giants of Garbage: The Rise of the Global Waste Industry and the Politics of Pollution Control* (Toronto: James Lorimer, 1993), 24.
85 "Probe's Last Word on 'Recycling,'" August 1978, 1, Projects/Reports/Submissions 1978, PPP.
86 Jack McGinnis, interview with author, 8 July 2008, conducted by telephone.
87 Ibid.

88 Ibid.; Humphries, *We Recycle*, 4.
89 John Marshall, "Metro Alchemists Turn Garbage into Gold," *Globe and Mail*, 1 March 1978, 5.
90 Jack McGinnis, interview with author, 8 July 2008, conducted by telephone.
91 Jacques Bendavid, "Environmental Group Awaits New Funding," Toronto *Star*, 1 September 1976, F3; Marshall, "Metro Alchemists Turn Garbage into Gold," 5.
92 Derek Stephenson, interview with author, 11 December 2009, conducted by telephone.
93 "Environment/Jobs: Conflict or Harmony," *Another Newsletter* 2, 4 (December 1977): 2.
94 *Investigation of the Feasibility of Increasing Corrugate Cardboard Recovery through Industrial and Commercial Source Separation in Ontario* (Toronto: Is Five Foundation/Resource Integration Systems, 1979), i, 1, 8.
95 *Description and Evaluation of the East York Recycling Model* (Toronto: Resource Integration Systems, 1979), 7.
96 Is Five Foundation, *Development and Demonstration of a Customized Truck for Collection of Glass, Metal and Paper Refuse* (Ottawa: Technical Services Branch, Environment Canada, 1983), 1.
97 Ibid., 1-2; "Government Policies Thwarting Recycling, Conference to Be Told," *Globe and Mail*, 1 June 1978, 3; Jack McGinnis, interview with author, 8 July 2008, conducted by telephone; Humphries, *We Recycle*, 6; Katharine Partridge, "RCO Celebrates 20 Years!" *RCO Update*, October 1998, 2-7; Eric Hellman, interview with author, 12 January 2010, conducted by telephone.
98 Humphries, ibid., 5; Resource Integration Systems, *At-Source Recovery of Waste Materials from CFB Borden: The Viability of At-Source Recovery in Small Communities, Executive Summary* (Toronto: Resource Integration Systems, 1979), 1.
99 Derek Stephenson, interview with author, 11 December 2009, conducted by telephone.
100 Resource Integration Systems, *At-Source Recovery of Waste Materials from CFB Borden*, 7, 33-34.

CHAPTER 6: BEYOND THE FIRST WAVE

1 Elizabeth D. Blum's *Love Canal Revisited* makes just one reference, in its endnotes, to Canadian groups and their involvement in the New York hazardous waste cases. *Love Canal Revisited: Race, Class, and Gender in Environmental Activism* (Lawrence: University Press of Kansas, 2008), 170n96.
2 Rae Tyson and Michael Keating, "Citizens' Groups Delay Plan to Clean U.S. Dump," *Globe and Mail*, 13 February 1982, 4; "Citizen's Groups Challenge Hooker Chemicals over Hyde Park Agreement," 11 May 1981, Press Releases, 1981, PPP.
3 Toby Vigod to John T. Curtin, US District Court, 7 May 1981, Hyde Park 1981, PPP.
4 Toby Vigod and Barbara Morrison, "Brief," 30 June 1981, Pollution Probe et al. vs. Hooker Chemical (Testimony) 1981, PPP; Rae Tyson, "Canadians to Review Waste Case," *Globe and Mail*, 30 May 1981, 4; Rae Tyson, "Dump Polluting Lake Ontario, Soil and Water Samples Confirm," *Globe and Mail*, 7 October 1981, 4; Blum, *Love Canal Revisited*, 113.
5 Stan Oziewicz, "No Appeal by Ontario on Dumping," *Globe and Mail*, 18 January 1980, 9; Anne Wordsworth and Brian Marshall to Keith Norton, 10 December 1981, Hyde Park 1981, PPP.
6 "Norton Refuses to Shun U.S. Case," *Globe and Mail*, 14 October 1982, 4.

7 Pollution Probe and Operation Clean – Niagara's request for intervener status was rejected because Judge John Curtin ruled that their interests were already represented by the government of Ontario. Jock Ferguson, "U.S. Ruling Irks Environmental Groups," *Globe and Mail*, 14 March 1983, M3.
8 Michael Keating, "Pollution Probe Assails Ontario's Dump Case as Hearings Adjourned," *Globe and Mail*, 5 May 1984, 21.
9 Michael Keating, "Brandt Denies Ontario Failed at Niagara Dump Hearings," *Globe and Mail*, 4 May 1984, M3; "Brandt Defends Role of S-Dump Delegates," *Globe and Mail*, 8 May 1984, M3.
10 Stan Oziewicz, "No Appeal by Ontario on Dumping," *Globe and Mail*, 18 January 1980, 9; "Waste Firm, Critics Agree," *Globe and Mail*, 15 April 1982, 3; Pollution Probe Foundation, "The Involvement of Environmental Interest Groups in the Development of the New York State Phased Ban on the Landfilling of Toxic Industrial Wastes: A Report," 2, Phased Ban on Landfilling Toxic Wastes 1984, PPP.
11 Quoted in Pollution Probe Foundation, ibid., 3.
12 Geoffrey York, "New York to Hold Hearings on Expansion of Dump Site," *Globe and Mail*, 4 November 1983, 17.
13 Pollution Probe Foundation, "The Involvement of Environmental Interest Groups," 5.
14 Michael Keating, "Wastes Policy Termed Threat to Water," *Globe and Mail*, 14 February 1984, 3.
15 Alden Baker, "Warning on Tap Water Was Wrong, MOH says," *Globe and Mail*, 11 November 1981, 5.
16 "Official Discounts Warning by Group over Water Quality," *Globe and Mail*, 17 February 1982, 5.
17 Michael Keating, "Tap Water Quality Criticized," *Globe and Mail*, 3 March 1983, 5.
18 Ibid.
19 "Bottled Water to Be Tested," *Globe and Mail*, 13 April 1983, 5; Michael Keating, "Advisers on Chemicals Named," *Globe and Mail*, 29 June 1983, 18; Michael Keating, "Brandt Awards Niagara Falls Pilot Project for Filtration," *Globe and Mail*, 3 April 1984, M3.
20 Pollution Probe's revival is discussed in David Lees, "Cleaning Up Our Act," *Globe and Mail*, 13 August 1983, F3. This is not to suggest, however, that Pollution Probe restricted its activities to these two areas during the period in question. For example, the organization maintained an active publishing program, including such popular works as: Linda Pim, *Additive Alert: A Guide to Food Additives for the Canadian Consumer* (Toronto: Doubleday Canada, 1979); Linda Pim, *The Invisible Additives: Environmental Contaminants in Our Food* (Toronto: Doubleday Canada, 1981); Monica Campbell and William Glenn, with assistance from Linda Pim, *Profit from Pollution Prevention: A Guide to Industrial Waste Reduction and Recycling* (Toronto: Pollution Probe Foundation, 1982); Pollution Probe Foundation, in consultation with Warner Troyer and Glenys Moss, *The Canadian Green Consumer Guide* (Toronto: McClelland and Stewart, 1989).
21 Colin Isaacs, interview with author, 18 August 2008, conducted by telephone.
22 Pollution Probe Foundation, "The Pollution Probe Foundation Annual Report 1978," 4, Annual Reports, PPP; Pollution Probe Foundation, "The Pollution Probe Foundation Annual Report 1981-1982," 9, Annual Reports, PPP; Pollution Probe Foundation, "The Pollution Probe Foundation Annual Report 1983," 10, Annual Reports, PPP; Judy Steed, "Opening Eyes to Pollution," *Globe and Mail*, 3 September 1988, D5.
23 Colin Isaacs, interview with author, 18 August 2008, conducted by telephone.
24 Ibid.; David Coon, interview with author, 18 August 2008, conducted by telephone.

25 Colin Isaacs, interview with author, 18 August 2008, conducted by telephone.
26 John Temple, "Pollution Probe Works a Deal with Masse," Toronto *Star*, 8 June 1988, A1.
27 Craig McInnes, "'Green' Products May Offer More for Conscience than Environment," *Globe and Mail*, 5 June 1989, A13; Lorne Slotnick, "Environmental Groups in Turmoil over Their Product Endorsements," *Globe and Mail*, 27 June 1989, A5; Lynda Hurst, "The Diapers Did Him In," Toronto *Star*, 8 July 1989, D4; Colin Isaacs, interview with author, 18 August 2008, conducted by telephone.
28 Clifford Maynes, "On Sports Star Level," *Globe and Mail*, 2 August 1989, A6.
29 Colin Isaacs, "Harnessing the Profit Motive to Clean Up World Pollution: It's Faster than Government," *Globe and Mail*, 10 July 1989, A7.
30 Untitled document, n.d., Consumer Packaging Survey, F1058 MU7337, AO.
31 As Isaacs explained: "It weakened the growing relationship between industry and the environmental community. I think that was unfortunate, but if I point the finger anywhere I point at the board for refusing to stand fast to a position they had already taken because the board was well aware that Pollution Probe was going to be endorsing products. They had agreed with it, but when the going got rough they sort of turned tail and ran." Colin Isaacs, interview with author, 18 August 2008, conducted by telephone.
32 Energy Probe Research Foundation 1983 Annual Report, quoted in Dewar, *Cloak of Green*, 454.
33 Patricia Adams and Lawrence Solomon, *In the Name of Progress: The Underside of Foreign Aid* (Toronto: Energy Probe Research Foundation, 1985); Stephen Dale, *McLuhan's Children: The Greenpeace Message and the Media* (Toronto: Between the Lines, 1996), 56-58.
34 Lawrence Solomon, interview with author, 14 August 2008, conducted by telephone.
35 Ibid. See also Dewar, *Cloak of Green*, 359-60.
36 Nyle Ludolph, interview with author, 16 January 2010, conducted by telephone.
37 Ibid. See also Humphries, *We Recycle*, 5-6.
38 Eric Hellman, interview with author, 12 January 2010, conducted by telephone.
39 Humphries, *We Recycle*, 7; Nyle Ludolph, interview with author, 16 January 2010, conducted by telephone.
40 Derek Stephenson, interview with author, 11 December 2009, conducted by telephone.
41 Humphries, *We Recycle*, 8.
42 Nyle Ludolph, interview with author, 16 January 2010, conducted by telephone.
43 Ibid.; Derek Stephenson, interview with author, 11 December 2009, conducted by telephone; Humphries, *We Recycle*, 8.
44 Canadian Institute for Environmental Law and Policy, "A Brief History of Waste Diversion in Ontario: A Background Paper on the Review of the *Waste Diversion Act*," November 2008, 2, http://www.cielap.org/pdf/WDA_BriefHistory.pdf; David McRobert, "Ontario's Blue Box System: A Case Study of Government's Role in the Technological Change Process, 1970-1991," LLM thesis, Osgoode Hall Law School, 1994, 40.
45 "The Story of Ontario's Blue Box," Stewardship Ontario, n.d., http://www.stewardshipontario.ca/wp-content/uploads/2013/02/Blue-Box-History-eBook-FINAL-022513.pdf; "Waste Diversion Ontario to redesign Ontario's Blue Box database," Waste Diversion Ontario, 7 May 2013, http://www.newswire.ca/en/story/1160335/waste-diversion-ontario-to-redesign-ontario-s-blue-box-database.
46 Derek Ferguson, "NDP Record on Refillables Criticized," Toronto *Star*, 24 November 1993, A10.

47 Doern and Conway, *The Greening of Canada*, 103; Robert Paehlke, "Canada," in *Capacity Building in National Environmental Policy: A Comparative Study of 17 Countries*, ed. Helmut Weidner and Martin Jänicke (New York: Springer, 2002), 128, 136-37.
48 Friends of the Earth Canada restructured in the 1990s, abandoning its umbrella group format in favour of direct membership.
49 Canadian Coalition on Acid Rain fonds, University of Waterloo Special Collections, http://www.lib.uwaterloo.ca/discipline/SpecColl/acid/; Monte Hummel, interview with author, 23 January 2008, Toronto; Mowat, *Rescue the Earth!* 34-46; Dewar, *Cloak of Green*, 333-35; Dan McDermott, interview with author, 26 June 2008, conducted by telephone; "History," Sierra Club Canada, http://www.sierraclub.ca/national/aboutus/history.html.
50 The Canadian Environmental Defence Fund was later renamed Environmental Defence Canada, while the Sierra Legal Defence Fund became Ecojustice Canada. Macdonald, *The Politics of Pollution*, 177; Peter Dauvergne, *Historical Dictionary of Environmentalism* (Lanham, MD: Scarecrow Press, 2009), 24; "Our Pembina Story," Pembina Institute, http://www.pembina.org/25Years; George Hoberg, "How the Way We Make Policy Governs the Policy We Make," in *Sustaining the Forests of the Pacific Coast: Forging Truces in the War in the Woods*, ed. Debra Salazar and Donald K. Alper (Vancouver: UBC Press, 2000), 34; David Suzuki, *David Suzuki: The Autobiography* (Vancouver: Greystone Books, 2006), 217-39; "Our Story," David Suzuki Foundation, http://www.davidsuzuki.org/about/our-story/.
51 *Rescue the Earth!* 34. As of 2014, Hummel remains with the WWFC as its president emeritus.
52 Campbell, Clément, and Kealey, *Debating Dissent*, 11.

Afterword

1 Ann Love, interview with author, 4 April 2008, conducted by telephone.
2 Peter Middleton, interview with author, 21 February 2008, Toronto. Further evidence of Pollution Probe's selective recognition was on display at the organization's thirtieth-anniversary gala celebration in 1999. Just five individuals were recognized for their contributions as founders: Sherry Brydson, Donald Chant, Tony Barrett, and Gage and Betty Love, who were hailed as Pollution Probe's "Godparents" on account of their numerous children and in-laws who became involved with the organization. "The Pollution Probe Thirtieth Anniversary Gala," booklet, Donald Chant papers, in the possession of Merle Chant. The need for working space in the often-cramped headquarters of a Canadian ENGO has been a major culprit in the loss of organizational memory.
3 Sponsors of the 2012 gala include the Cement Association of Canada, the Imperial Oil Foundation, Procter and Gamble Canada, Royal Bank of Canada, Shell Canada, Suncor Energy, Union Gas, and Vale Canada. "Pollution Probe's 2013 Annual Gala Dinner," http://www.pollutionprobe.org/gala.
4 The notable exception to this is Pollution Probe's annual Clean Air Commute, a heavily advertised event that encourages residents of the Greater Toronto Area to "walk, bike, telecommute or take transit instead of driving to work." This weeklong event prevented 380 tonnes of pollutants from entering the atmosphere in 2013. "Clean Air Commute," http://www.cleanaircommute.ca.

Bibliography

ARCHIVAL SOURCES

Archives of Ontario, Toronto
Correspondence of the Deputy Minister of Health
Energy Probe fonds
Pesticides Advisory Committee files
Pollution Probe fonds

Canadian Broadcasting Centre, Toronto
Canadian Broadcasting Corporation Reference Library

City of Toronto Archives
Tony O'Donohue fonds

Library and Archives Canada, Ottawa
Canadian Broadcasting Corporation fonds

University of Toronto Archives
Office of the President fonds
Omond M. Solandt fonds

University of Waterloo Special Collections
Canadian Coalition on Acid Rain fonds

Wilfrid Laurier University Archives, Waterloo, ON
John Swaigen fonds

Bibliography

Private Collections

Donald Chant papers, in the possession of Merle Chant, Madoc, ON
Larry Gosnell papers, in the possession of Denise Gosnell, Toronto
Pollution Probe papers, located at Pollution Probe headquarters, Toronto

Written Correspondence

Conway, Chris, 9 September 2010
Middleton, Peter, 19 July 2010
Power, Dennis, 21 July 2010

Newspapers and Newsletters

Another Newsletter, 1977
Environmental Law News, 1972-73
Family Herald, 1967
Globe and Mail, 1967-89
Hamilton *Spectator*, 1969
Industrial Canada, 1968
Ottawa *Citizen*, 1969-71
Probe Post, 1980
RCO Update, 1998
Toronto *Star*, 1966-88
Toronto *Telegram*, 1967-71
University of Toronto *Varsity*, 1969-77

Reports

Advisory Committee on Energy. *Energy in Ontario: The Outlook and Policy Implications, Volume 1*. Toronto: Ontario Legislative Assembly Committee on Government Productivity, 1973.
Berger, Thomas. *Northern Frontier, Northern Homeland: The Report of the Mackenzie Valley Pipeline Inquiry.* Toronto: Douglas and McIntyre, 1988.
Canada as a Conserver Society: An Agenda for Action. Ottawa: Science Council of Canada, 1978.
"The Canada's World Poll." Environics Research Group and the Simons Foundation, January 2008. http://www.environicsinstitute.org/uploads/institute-projects/the%20canada%27s%20world%20poll%20-%20final%20report.pdf.
Description and Evaluation of the East York Recycling Model. Toronto: Resource Integration Systems, 1979.
Edwards, Martin H. *Did Pesticides Kill Ducks on Toronto Island? Report of the Royal Commission Appointed to Inquire into the Use of Pesticides and the Death of Waterfowl on Toronto Island.* Toronto: Queen's Printer, 1970.

General Report of the Solid Waste Task Force to the Minister of the Environment, Part 1. Toronto: Ministry of the Environment, 1974.

Investigation of the Feasibility of Increasing Corrugate Cardboard Recovery through Industrial and Commercial Source Separation in Ontario. Toronto: Is Five Foundation/Resource Integration Systems, 1979.

Is Five Foundation. *Development and Demonstration of a Customized Truck for Collection of Glass, Metal and Paper Refuse.* Ottawa: Technical Services Branch, Environment Canada, 1983.

"Low income measures, low income after-tax cut-offs and low income after tax measures." Statistics Canada. 7 April 2004. http://www.statcan.gc.ca/pub/13f0019x/13f0019x 1997000-eng.pdf.

Ontario Royal Commission on Fluoridation. *Report of the Committee Appointed to Inquire into and Report upon Fluoridation of Municipal Water Supplies.* Toronto: Queen's Printer, 1961.

Pollution Inquiry Committee. *Report of the Committee Appointed to Inquire into and Report upon the Pollution of Air, Soil, and Water in the Townships of Dunn, Moulton, and Sherbrooke, Haldimand County.* Toronto: Queen's Printer, 1968.

Resource Integration Systems. *At-Source Recovery of Waste Materials from CFB Borden: The Viability of At-Source Recovery in Small Communities – Executive Summary.* Toronto: Resource Integration Systems, 1979.

"The Story of Ontario's Blue Box." Stewardship Ontario, n.d. http://www.stewardship ontario.ca/wp-content/uploads/2013/02/Blue-Box-History-eBook-FINAL-022513.pdf.

Task Force Hydro. *Hydro in Ontario: A Future Role and Place.* Toronto: Ontario Legislative Assembly Committee on Government Productivity, 1973.

–. *Hydro in Ontario: An Approach to Organization.* Toronto: Ontario Legislative Assembly Committee on Government Productivity, 1972.

–. *Hydro in Ontario: Financial Policy and Rates.* Toronto: Ontario Legislative Assembly Committee on Government Productivity, 1973.

–. *Nuclear Power in Ontario.* Toronto: Ontario Legislative Assembly Committee on Government Productivity, 1973.

Audio-Visual Sources

The Air of Death. Directed by Larry Gosnell. 1967. Toronto: CBC Archive Sales, 2008. DVD.

"Dishing the Dirt on Phosphates." CBC Digital Archives. Originally broadcast on 8 February 1970. http://archives.cbc.ca/environment/pollution/topics/1390/.

"Greenpeace Protests at Darlington Site." CBC Digital Archives. Originally broadcast on 1 October 1977. http://archives.cbc.ca/science_technology/energy_production/clips/915/.

Interviews

Aarons, Marilyn, telephone, 29 November 2009
Alden, Terry, telephone, 19 January 2010
Argue, Bob, telephone, 8 December 2009

Bibliography

Attwater, Clive, telephone, 3 October 2008
Bacque, James, 24 March 2008
Bennett, John, telephone, 8 July 2008
Bielfeld, Arthur, telephone, 21 December 2009
Brinkhurst, Ralph, telephone, 8 August 2008
Brooks, David, telephone, 13 July 2009
Bryce, Gregory, telephone, 18 March 2008
Burke, Stanley, telephone, 2 November 2007
Burke, Stanley, 25 January 2008
Chant, Donald, telephone, 18 November 2007
Chant, Merle, 25 January 2008
Conway, Chris, telephone, 14 July 2009
Coombs, John, telephone, 13 November 2008
Coon, David, telephone, 18 August 2008
Creed, Murray, telephone, 28 January 2008
Davis, Bill, telephone, 5 June 2008
Edwards, Martin H., telephone, 2 January 2008
Estrin, David, 19 March 2008
Foster, Reg, telephone, 3 October 2008
Gibson, Robert, telephone, 16 October 2009
Gosnell, Denise, 19 March 2008
Hellman, Eric, telephone, 12 January 2010
Hummel, Monte, 23 January 2008
Isaacs, Colin, telephone, 18 August 2008
Kelly, Brian, telephone, 12 January 2009
Levy, Alan, 19 March 2008
Love, Ann, telephone, 4 April 2008
Love, Geoff, telephone, 11 July 2008
Love, Peter, 10 March 2008
Ludolph, Nyle, telephone, 16 January 2010
Mahood, Gar, 4 March 2008
McDermott, Dan, telephone, 26 June 2008
McGinnis, Jack, telephone, 8 July 2008
McQuail, Tony, telephone, 19 August 2008
McQuay, Jan (Jan Marmorek), telephone, 23 October 2009
Middleton, Peter, 21 February 2008
Mills, Rob, telephone, 25 September 2008
O'Donohue, Tony, 10 October 2007
O'Malley, Terry, telephone, 8 July 2008
Opperman, Jo, telephone, 3 February 2010
Osler, Sanford, telephone, 12 August 2008
Paehlke, Robert, telephone, 19 April 2010
Palmer, Janice, telephone, 2 March 2010
Pitman, Walter, telephone, 6 January 2010
Plowright, Chris, telephone, 30 January 2008
Poch, David, telephone, 6 January 2010
Regier, Henry, telephone, 14 August 2008

Rounthwaite, Ann, telephone, 10 July 2008
Saunders, Douglas, telephone, 9 August 2008
Schwass, Rodger, telephone, 19 April 2008
Sewell, John, 5 March 2008
Sheard, Gloria, 24 March 2008
Sheard, Joseph, 24 March 2008
Sirluck, Ernest, 7 November 2007
Solomon, Lawrence, telephone, 14 August 2008
Spink, Lynn (Marilyn Cox), 12 March 2008
Spinner, Barry, telephone, 24 July 2009
Stephenson, Derek, telephone, 11 December 2009
Ware, Meredith, telephone, 2 December 2008
Warwick, Joe, telephone, 6 August 2008
Willms, John, 24 April 2008
Wood, David, telephone, 20 March 2008
Wright, Janet, telephone, 27 October 2009
Zlotkin, Stanley, 19 February 2008

SECONDARY SOURCES

Adams, Patricia, and Lawrence Solomon. *In the Name of Progress: The Underside of Foreign Aid*. Toronto: Energy Probe Research Foundation, 1985.
Anastakis, Dimitry, ed. *The Sixties: Passion, Politics, and Style*. Montreal and Kingston: McGill-Queen's University Press, 2008.
Ayre, John. "A Student Puts the Case for Reform as Our New Campus Activists See It." University of Toronto *Graduate* 3, 4 (January 1971): 16.
Babin, Ronald. *The Nuclear Power Game*. Montreal: Black Rose Books, 1985.
Banting, Keith, George Hoberg, and Richard Simeon, eds. *Degrees of Freedom: Canada and the United States in a Changing World*. Toronto: University of Toronto Press, 1997.
Bera, Yogi. *The Yogi Book*. New York: Workman Publishing, 1998.
Berman, Tzeporah (with Mark Leiren-Young). *This Crazy Time: Living Our Environmental Challenge*. Toronto: Alfred A. Knopf Canada, 2011.
Bissell, Claude. *Halfway Up Parnassus: A Personal Account of the University of Toronto, 1932-1971*. Toronto: University of Toronto Press, 1974.
Blum, Elizabeth D. *Love Canal Revisited: Race, Class, and Gender in Environmental Activism*. Lawrence: University Press of Kansas, 2008.
Bosso, Christopher J. *Environment, Inc.: From Grassroots to Beltway*. Lawrence: University Press of Kansas, 2005.
Bothwell, Robert. *Alliance and Illusion: Canada and the World, 1945-1984*. Vancouver: UBC Press, 2007.
Braunstein, Peter, and Michael William Doyle, eds. *Imagine Nation: The American Counterculture of the 1960s and '70s*. New York: Routledge, 2002.
Brooks, David. *Zero Energy Growth for Canada*. Toronto: McClelland and Stewart, 1981.
Bullough, Vern L., ed. *Encyclopedia of Birth Control*. Santa Barbara, CA: ABC-CLIO, 2001.

Burgess, Diane. "*Kanehsatake* on *Witness:* The Evolution of CBC Balance Policy." *Canadian Journal of Communication* 25, 2 (2000): n.p. http://www.cjc-online.ca/index.php/journal/article/viewArticle/1153/1072.

Burgstahler, Albert W. "George L. Waldbott – A Pre-Eminent Leader in Fluoride Research." *Fluoride* 31, 1 (February 1998): 2-4.

Campbell, Lara, Dominique Clément, and Gregory S. Kealey, eds. *Debating Dissent: Canada and the Sixties*. Toronto: University of Toronto Press, 2012.

Campbell, Monica, and William Glenn. *Profit from Pollution Prevention: A Guide to Industrial Waste Reduction and Recycling*. With the assistance of Linda Pim. Toronto: Pollution Probe Foundation, 1982.

"Canada Pulls Out of Kyoto Protocol." CBC News, 12 December 2011. http://www.cbc.ca/news/politics/story/2011/12/12/pol-kent-kyoto-pullout.html.

Canadian Institute for Environmental Law and Policy. "A Brief History of Waste Diversion in Ontario: A Background Paper on the Review of the *Waste Diversion Act*." November 2008. http://www.cielap.org/pdf/WDA_BriefHistory.pdf.

"Canadians Continue to Voice Strong Support for Actions to Address Climate Change, Including an International Treaty and Carbon Taxes." Environics Research Group, 1 December 2011. http://www.environics.ca/reference-library?news_id=109.

Captain Enviro: The Fight to Save the Maritimes! http://www.misterkitty.org/extras/stupidcovers/stupidcomics147.html.

Carroll, William K., ed. *Organizing Dissent: Contemporary Social Movements in Theory and Practice*. 2nd ed. Toronto: Garamond Press, 1997.

Carstairs, Catherine, and Rachel Elder. "Expertise, Health, and Popular Opinion: Debating Water Fluoridation, 1945-1980." *Canadian Historical Review* 89, 3 (September 2008): 345-71.

A Celebration of the Life of Dr. Donald Chant. Booklet published by family, 2008.

"Clean Air Commute." http://www.cleanaircommute.ca.

Clément, Dominique. *Canada's Rights Revolution: Social Movements and Social Change, 1937-1982*. Vancouver: UBC Press, 2008.

Cook, Peter G., and Myles A. Ruggles. "Balance and Freedom of Speech: Challenge for Canadian Broadcasting." *Canadian Journal of Communication* 17, 1 (1992): n.p. http://www.cjc-online.ca/index.php/journal/article/viewArticle/647/553#.

Coudreau, Julie-Ann, Roger Keil, and Douglas Young. *Changing Toronto: Governing Urban Neoliberalism*. Toronto: University of Toronto Press, 2009.

Crooks, Harold. *Giants of Garbage: The Rise of the Global Waste Industry and the Politics of Pollution Control*. Toronto: James Lorimer, 1993.

Dale, Stephen. *McLuhan's Children: The Greenpeace Message and the Media*. Toronto: Between the Lines, 1996.

Dauvergne, Peter. *Historical Dictionary of Environmentalism*. Lanham, MD: Scarecrow Press, 2009.

Dewar, Elaine. *Cloak of Green*. Toronto: James Lorimer, 1995.

Doern, G. Bruce, and Thomas Conway. *The Greening of Canada: Federal Institutions and Decisions*. Toronto: University of Toronto Press, 1994.

Doern, G. Bruce, and Stephen Wilks, eds. *Changing Regulatory Institutions in Britain and North America*. Toronto: University of Toronto Press, 1998.

Dowie, Mark. *Losing Ground: American Environmentalism at the Close of the Twentieth Century*. Cambridge, MA: The MIT Press, 1995.

Dubinsky, Karen, Catherine Krull, Susan Lord, Sean Mills, and Scott Rutherford, eds. *New World Coming: The Sixties and the Shaping of Global Consciousness.* Toronto: Between the Lines, 2009.

Dunlap, Riley E., and Angela G. Mertig, eds. *American Environmentalism: The U.S. Environmental Movement, 1970-1990.* Philadelphia: Taylor and Francis, 1992.

Dunlap, Thomas R. *DDT: Scientists, Citizens, and Public Policy.* Princeton, NJ: Princeton University Press, 1981.

Ehrlich, Paul. *The Population Bomb.* New York: Ballantine Books, 1968.

Elton, Sarah. "Green Power." *U of T Magazine* (Winter 1999). http://www.magazine.utoronto.ca/feature/canadian-environmental-movement/.

Emond, D. Paul. "'Are We There Yet?' Reflections on the Success of the Environmental Law Movement in Ontario." *Osgoode Hall Law Journal* 46, 2 (Summer 2008): 219-42.

"Environment a Priority for More Canadians, Poll Suggests." CBC News, 8 November 2006. http://www.cbc.ca/news/canada/story/2006/11/08/environment-poll.html.

Finlayson, Ann. "Open the House, Close the Door." *Maclean's,* 24 November 1980, np.

Fitzpatrick, Meagan. "Environment Canada Job Cuts Raise Concerns." CBC News, 4 August 2011. http://www.cbc.ca/news/politics/story/2011/08/04/pol-environment-job-cuts.html.

Ford, Ray. "Death and the Rebirth of the Don River." *Canadian Geographic,* June 2011. http://www.canadiangeographic.ca/magazine/june11/don_river_watershed.asp.

Forkey, Neil S. *Canadians and the Natural Environment to the Twenty-First Century.* Toronto: University of Toronto Press, 2012.

Freeman, Neil B. *The Politics of Power: Ontario Hydro and Its Government, 1906-1995.* Toronto: University of Toronto Press, 1996.

Friedland, Martin L. *The University of Toronto: A History.* Toronto: University of Toronto Press, 2002.

Friel, Brian. "Faith Healer." In *Brian Friel: Plays One,* 327-76. London: Faber and Faber, 1996.

Frisken, Frances. *The Public Metropolis: The Political Dynamics of Urban Expansion in the Toronto Region, 1924-2003.* Toronto: Canadian Scholars' Press, 2007.

Frizzell, Alan, and John H. Pammett, eds. *Shades of Green: Environmental Attitudes in Canada and around the World.* Montreal and Kingston: McGill-Queen's University Press, 1997.

Gore, Al. *Inconvenient Truth: The Planetary Emergency of Global Warming and What We Can Do About It.* Emmaus, PA: Rodale Press, 2006.

Gottlieb, Robert. *Forcing the Spring: The Transformation of the American Environmental Movement.* Washington, DC: Island Press, 2005.

Graham, Otis L. Jr., ed. *Environmental Politics and Policy, 1960s-1990s.* University Park: Pennsylvania State University Press, 2000.

Hagan, John. *Northern Passage: American Vietnam War Resisters in Canada.* Cambridge, MA: Harvard University Press, 2001.

Haley, Ella. "Methodology to Deconstruct Environmental Inquiries Using the Hall Commission as a Case Study." PhD dissertation, York University, 2000.

Hays, Samuel P. *Explorations in Environmental History: Essays.* Pittsburgh, PA: University of Pittsburgh Press, 1998.

Henderson, Stuart. *Making the Scene: Yorkville and Hip Toronto in the 1960s.* Toronto: University of Toronto Press, 2011.

"History." Sierra Club Canada. http://www.sierraclub.ca/national/aboutus/history.html.

"How Mrs. Boston Burst a Great Detergent Bubble." *Maclean's,* April 1970, 3.

Humphries, Diane. *We Recycle: The Creators of the Blue Box Programme.* Toronto: Pollution Probe, 1997.

"Ian Efford/Brian Kelly: The Panjandrums of Zero Growth." *Saturday Night,* March 1977, 34-35.

Jarrett, Henry, ed. *Perspectives on Conservation.* Baltimore: Johns Hopkins University Press, 1958.

Keeling, Arn. "Urban Waste Sinks as a Natural Resource: The Case of the Fraser River." *Urban History Review* 34, 1 (Fall 2005): 58-70.

Kinsman, Gary, and Patrizia Gentile. *The Canadian War on Queers: National Security as Sexual Regulation.* Vancouver: UBC Press, 2009.

Kostash, Myrna. *Long Way from Home: The Story of the Sixties Generation in Canada.* Toronto: James Lorimer, 1980.

Laxer, James. *The Energy Poker Game.* Toronto: New Press, 1970.

Leiss, William, ed. *Ecology versus Politics in Canada.* Toronto: University of Toronto Press, 1979.

Lemon, James. *Toronto since 1918: An Illustrated History.* Toronto: James Lorimer, 1985.

Levitt, Cyril. *Children of Privilege: Student Revolt in the Sixties.* Toronto: University of Toronto Press, 1984.

Levy, Alan. "Readers Digest of CELA's History." *Intervenor* 26, 1 (January-March 2001). http://www.cela.ca/article/readers-digest-celas-history.

Lexier, Roberta. "The Canadian Student Movement in the Sixties: Three Case Studies." PhD dissertation, University of Alberta, 2009.

Macdonald, Doug. *The Politics of Pollution.* Toronto: McClelland and Stewart, 1991.

Macdonald, Douglas. *Business and Environmental Politics in Canada.* Peterborough, ON: Broadview Press, 2007.

MacEachern, Alan. *The Institute of Man and Resources: An Environmental Fable.* Charlottetown: Island Studies Press, 2003.

McCarthy, John D., and Mayer N. Zald. "Resources Mobilization and Social Movements: A Partial Theory." *American Journal of Sociology* 82, 6 (May 1977): 1212-41.

McCormick, John. *Reclaiming Paradise: The Global Environmental Movement.* Bloomington and Indianapolis: Indiana University Press, 1989.

McKenzie, Judith I. *Environmental Politics in Canada: Managing the Commons into the Twenty-First Century.* Toronto: Oxford University Press, 2002.

McLaughlin, Mark J. "Green Shoots: Aerial Insecticide Spraying and the Growth of Environmental Consciousness in New Brunswick, 1952-1973." *Acadiensis* 40, 1 (Winter/Spring 2011): 3-23.

–. "Rise of the Eco-Comics: The State, Environmental Education and Canadian Comic Books, 1971-1975." *Material Culture Review/Revue de la culture matérielle* 77, 78 (Spring/Fall 2013): 9-20.

McRobert, David. "Ontario's Blue Box System: A Case Study of Government's Role in the Technological Change Process, 1970-1991." LLM thesis, Osgoode Hall Law School, 1994.

Mehta, Michael D. *Risky Business: Nuclear Power and Public Protest in Canada.* Toronto: Lexington Books, 2005.

Melosi, Martin V. *Garbage in the Cities: Refuse, Reform, and the Environment.* Pittsburgh, PA: University of Pittsburgh Press, 2005.

Merchant, Livingston T., and A.D.P. Heeney. "Canada and the United States: Principles for Partnership." *Atlantic Community Quarterly* 3, 3 (Fall 1965): 373-91.
Mills, Sean. *The Empire Within: Postcolonial Thought and Political Activism in Sixties Montreal.* Montreal and Kingston: McGill-Queen's University Press, 2010.
Mowat, Farley. *Rescue the Earth! Conversations with the Green Crusaders.* Toronto: McClelland and Stewart, 1990.
O'Connor, Ryan. "Advertising the Environmental Movement: Vickers and Benson's Branding of Pollution Probe." *Rachel Carson Center Perspectives* 1 (2013). http://www.environmentandsociety.org/sites/default/files/seiten_aus_2013_11_web_final2_kleiner-3_0.pdf.
"Our Funding." Energy Probe Research Foundation. http://epresearchfoundation.wordpress.com/our-funding/.
"Our Pembina Story." Pembina Institute. http://www.pembina.org/25Years.
"Our Story." David Suzuki Foundation. http://www.davidsuzuki.org/about/our-story/.
Owram, Doug. *Born at the Right Time: A History of the Baby Boom Generation.* Toronto: University of Toronto Press, 1996.
Paehlke, Robert. "Eco-History: Two Waves in the Evolution of Environmentalism." *Alternatives* 19, 1 (September 1992): 18-23.
Page, Robert. *Northern Development: The Canadian Dilemma.* Toronto: McClelland and Stewart, 1986.
Palaeologu, M. Athena, ed. *The Sixties in Canada: A Turbulent and Creative Decade.* Montreal: Black Rose Books, 2009.
Palmer, Bryan. *Canada's 1960s: The Ironies of Identity in a Rebellious Era.* Toronto: University of Toronto Press, 2009.
Patterson, Walt. "Porter with Nuclear Reservations." *New Scientist,* 2 November 1978, 362-63.
Payton, Laura. "Radicals Working against Oilsands, Ottawa Says." CBC News, 9 January 2012. http://www.cbc.ca/news/politics/story/2012/01/09/pol-joe-oliver-radical-groups.html.
"Peter Middleton: The Entrepreneur of Renewables." *Saturday Night,* March 1977, 32-33.
Phillips, Mark Salber. "Reconsidering Historical Distance." http://www.ugent.be/paradoctoralschools/en/doctoraltraining/courses/archive/2010-2011/2010-2011-reconsidering-historical-distance.htm.
—. "Rethinking Historical Distance: From Doctrine to Heuristic." *History and Theory* 50, 4 (December 2011): 12.
Pim, Linda. *Additive Alert: A Guide to Food Additives for the Canadian Consumer.* Toronto: Doubleday Canada, 1979.
—. *The Invisible Additives: Environmental Contaminants in Our Food.* Toronto: Doubleday Canada, 1981.
Pollution Probe Foundation. *The Canadian Green Consumer Guide.* In consultation with Warner Troyer and Glenys Moss. Toronto: McClelland and Stewart, 1989.
"Pollution Probe's 2013 Annual Gala Dinner." http://www.pollutionprobe.org/gala.
Pross, A. Paul, ed. *Pressure Group Behaviour in Canadian Politics.* Toronto: McGraw-Hill Ryerson, 1975.
Randall, Stephen J. *Foreign Oil Policy since World War I: For Profits and Security,* 2nd ed. Montreal and Kingston: McGill-Queen's University Press, 2005.

Read, Jennifer. "'Let Us Heed the Voice of Youth': Laundry Detergents, Phosphates and the Emergence of the Environmental Movement in Ontario." *Journal of the Canadian Historical Association* 7 (1996): 227-50.
Regier, Henry, and J. Bruce Falls, eds. *Exploding Humanity: The Crisis of Numbers*. Toronto: House of Anansi Press, 1969.
Richardson, Boyce. "Pollution: Everybody's Business." Montreal: Montreal *Star*, 1969.
Richardson, Jeremy J., ed. *Pressure Groups*. New York: Oxford University Press, 1993.
Robinson, John B., ed. *Life in 2030: Exploring a Sustainable Future for Canada*. Vancouver: UBC Press, 1996.
Rothschild, Joyce, and J. Allen Whitt. *The Cooperative Workplace: Potentials and Dilemmas of Organisational Democracy and Participation*. New York: Cambridge University Press, 1986.
Salazar, Debra, and Donald K. Alper, eds. *Sustaining the Forests of the Pacific Coast: Forging Truces in the War in the Woods*. Vancouver: UBC Press, 2000.
Sale, Peter. *Our Dying Planet: An Ecologist's View of the Crisis We Face*. Berkeley: University of California Press, 2011.
Schultz, George P., and Kenneth W. Dam. *Economic Policy beyond the Headlines*. Chicago: University of Chicago Press, 1998.
Siggins, Maggie. *Bassett*. Toronto: James Lorimer, 1979.
Solomon, Lawrence. *The Conserver Solution*. Toronto: Doubleday Canada, 1978.
–. "Letter: Foreign Donations OK: Suzuki Foundation." *Financial Post*, 19 December 2013. http://opinion.financialpost.com/2013/12/19/letter-foreign-donations-ok-suzuki-foundation/.
Speth, James Gustave. *The Bridge at the End of the World: Capitalism, the Environment, and Crossing from Crisis to Sustainability*. New Haven, CT: Yale University Press, 2008.
–. *Red Sky at Morning: America and the Crisis of the Global Environment*. New Haven, CT: Yale University Press, 2004.
St-Pierre, Marc. "Footprints: Environment and the Way We Live." National Film Board. N.d. http://www3.nfb.ca/footprints/nfb-and-environment/the-early-years.html?part=3.
Suzuki, David. *David Suzuki: The Autobiography*. Vancouver: Greystone Books, 2006.
Swift, Jamie, and Keith Stewart. *Hydro: The Decline and Fall of Ontario's Electric Empire*. Toronto: Between the Lines, 2004.
Tjornbo, Ola, Frances Westley, and Darcy Riddell (Social Innovation Generation @ University of Waterloo). Case Study 003. "The Great Bear Rainforest Story." January 2010. http://sig.uwaterloo.ca/highlight/case-study-the-great-bear-rainforest-story.
Van Huizen, Philip. "Building a Green Dam: Environmental Modernism and the Canadian-American Libby Dam Project." *Pacific Historical Review* 79, 3 (2010): 418-53.
–. "'Panic Park': Environmental Protest and the Politics of Parks in British Columbia's Skagit Valley." *BC Studies* 170 (Summer 2011): 67-92.
Wall, Sharon. *The Nurture of Nature: Childhood, Antimodernism, and Ontario Summer Camps, 1920-55*. Vancouver: UBC Press, 2009.
Walton, K.C. "Environmental Fluoride and Fluorosis in Mammals." *Mammal Review* 18, 2 (June 1988): 77-90.
Warecki, George M. *Protecting Ontario's Wilderness: A History of Changing Ideas and Preservation Politics, 1927-1973*. New York: Peter Lang Publishing, 2000.

"Waste Diversion Ontario to redesign Ontario's Blue Box database." Waste Diversion Ontario. 7 May 2013. http://www.newswire.ca/en/story/1160335/waste-diversion-ontario-to-redesign-ontario-s-blue-box-database.

Webb, Margaret. "The Age of Dissent." *U of T Magazine* (Spring 2002). http://www.magazine.utoronto.ca/feature/the-age-of-dissent/.

Weidner, Helmut, and Martin Jänicke, eds. *Capacity Building in National Environmental Policy: A Comparative Study of 17 Countries*. New York: Springer, 2002.

Weyler, Rex. *Greenpeace: How a Group of Ecologists, Journalists and Visionaries Changed the World*. Vancouver: Raincoast Books, 2004.

Winfield, Mark S. *Blue-Green Province: The Environment and the Political Economy of Ontario*. Vancouver: UBC Press, 2012.

"Wisest Fools." *The Economist*. 27 January 2005. http://www.economist.com/node/3598744.

Wordsworth, William. "The French Revolution as It Appeared to Enthusiasts at Its Commencement." *The Complete Poetical Works, by William Wordsworth ... With an Introduction by John Morley*. London: Macmillan, 1888.

Worster, Donald. *Nature's Economy: A History of Ecological Ideas*, 2nd ed. Cambridge: Cambridge University Press, 1994.

Wynn, Graeme. "Rethinking Environmentalism." Foreword to *Tracking the Great Bear: How Environmentalists Recreated British Columbia's Coastal Rainforest*. Vancouver: UBC Press, 2014, and this volume in toto.

Zelko, Frank. *Make It a Green Peace! The Rise of Countercultural Environmentalism*. Toronto: Oxford University Press, 2013.

–. "Making Greenpeace: The Development of Direct Action Environmentalism in British Columbia." *BC Studies* 142/143 (Summer/Autumn 2004): 197-239.

Index

Note: Page numbers in bold refer to figures

Aarons, Marilyn, 143, 145, **146**
Adams, Pat, **146**
Adamson, Harold, 49, 86-87
advertisements: against air pollution, 33, 57-58; *pro bono*, 51, 52; targetting Ontario Hydro, 57-58; against water pollution, 53
Advisory Committee on Energy (ACE), 107; final report, 108
The Air of Death: broadcast, 18-21; CRTC hearing, 6, 27-29, 185n55; CRTC report on, 29-30; development of, 13-18; and environmental movement in Toronto, 30-36; Hall Commission inquiry into, 23-26; praise for, 27; response to, 21-23; support for, xiii, xiv, 33-34, 184n36, 186n83
air pollution, xiii, xvi, 6, 18-21, 34, 56-59, 189n58; cost of, 131; cost of clean-up, 31
Air Pollution Control Act, 118
Air Pollution Control Bureau, 16
Alden, Terry, 61, 77
Allen, William, 47
Alleyn, Jacques, 28
alpha-chloralose, 45, 48
Alternatives, 66

Amchitka, 100-1
American Society for Fluoride Research, 17
Anderson, Grant, 155
Anderson, J.M., 27
anti-litter campaign, 48-50, 188n34, 188n37
anti-nuclear protests, 136, 139-40, 141
Arctic Waters Pollution Prevention Act, 109
Atomic Energy of Canada Ltd. (AECL), 125, 135
Attwater, Clive, 85, 87
Atyeo, Harold, 87, 88
Audubon Society, 3
Auld, James, 96-97, 110, 113, 118

Bacque, James, 30, 31, 32, 33, 78, 185n72
Baetz, Reuben, 138
Bales, Dalton, 94
Barnard, Dick, 57
Barnett, Henry, 101
Barrett, Tony, vi, 61; and anti-diazinon campaign, 39, 47; and anti-litter campaign, 49; eco-financier, 62, 65, 76-77; and free advertising space,

219

51-52; influence on Pollution Probe, 42-43; leaves Pollution Probe, 127; and Pierre Trudeau, 189n73; 207n2; Pollution Probe VP (Administration), 41; and Procter and Gamble, 64-65; professional environmentalist, 62; and public interest law clinic, 92; recognition as Pollution Probe founder, 175, 207n2; and Resources Recycling Caravan, 62; as Simon Greed, 54, 55
Bassett, John, 52
Baum, Gregory, 70
Beckett, Thomas, 57
Becking, George, 158
Bennett, John, 134, 135, 136, 140
benzene, 158
benzopyrene, xiii, 18
Berger, Thomas, 110
Berman, Tzeporah, xxi
Bernhart, Alfred, 30, 31, 32, 33-34, 46
Berton, Pierre, 100
Biggs, Everett, 16
Bissell, Claude, 44, 64, 78
Blackburn, Bob, 22
blue boxes, 164-67, 166
Blum, Elizabeth, 155
Board of Control, York Council, 118
Boorsma, Ted, 19, 20
Bosso, Christopher, 3, 12
Bowle, John, 64-65
Boyle, Harry J., 28
Brandt, Andrew, 156
Bray, A.S., 197n23
Breaking Up Ontario Hydro's Monopoly (Solomon), 146
Brinkhurst, Ralph, 37, 42, 43, 61, 71, 189n73
British Columbia Recycling Council, 151
British Columbia Sportsmen's Council, 3
Brooks, David, 132-33, 145, 146, 203n80
Brown, R.D., 76
Bruce Nuclear Generating Station, 136
Bryce, Gregory, 85, 87, 89, 110, 113, 116, 172; recognition as Pollution Probe founder, 175
Brydson, Sherry, xiv, xviii-xix, 42, 186n74, 186n84, 207n2; and founding of Pollution Probe, 34-35, 36; recognition as Pollution Probe founder, 175, 207n2
Burke, Stanley, 20; and *The Air of Death*, xiii, 6, 12, 18-21; background, 14; and Toronto environmental movement, 27, 28-29, 30-31, 37, 43, 78, 185n61, 185n71
Burstyn, Varda, 10

Cadbury, Barbara, 70
Cadbury, George, 70
Campbell, Lara, 9, 170
Campbell, Margaret, 47
Canada: birth rate, 71; domestic nuclear power program, 125, 126; energy pact with US, 80-84; energy policy, 59, 81-82, 83-84; environmental policy, 48, 95, 96, 106-7, 118, 128, 166-67
Canada as a Conserver Society: An Agenda for Action, 129-30
Canada Water Act, 60
Canadian Arctic Gas, 109
Canadian Arctic Resources Committee, 110
Canadian Association on the Human Environment (CAHE), 75, 91-92
Canadian Broadcasting Corporation (CBC), 56; account of Bruce Nuclear Generating Station security breach, 136; and *The Air of Death*, 6, 184n36; broadcast of Energy Project's agenda, 82; and CRTC hearing, 27-29; coverage of Don River funeral, 56; and Hall Commission, 25, 26; segment on phosphates, 60
Canadian Campaign for Nuclear Disarmament, 125
Canadian Coalition on Acid Rain (CCAR), 4, 8, 168
Canadian Coalition for Nuclear Responsibility, 138
Canadian Council of Resource Ministers conference, 13, 14
Canadian Dental Association, 17
Canadian Environmental Defence Fund, 169. *See* Environmental Defence Canada
Canadian Environmental Law Association (CELA), 5, 7, 154, 169; and bipartite

bargaining system, 172; and Ontario government, 95-97; purpose, 76; and Sandbanks Park cases, 93-96
Canadian Environmental Law Research Foundation (CELRF), 5, 7, 95-96, 194n49
Canadian Forces Base Borden, 151
Canadian Gypsum, 117-18, 199n58
Canadian International Development Agency, 162-63
The Canadian Junior Green Guide, 174-75
Canadian Medical Association, 17
Canadian Petroleum Association, 145
Canadian Radio-Television Commission (CRTC): *The Air of Death* hearing, 6, 27-29, 185n55; report, 29-30
Canadian Wildlife Service, 155
CANDU: An Analysis of the Canadian Nuclear Program, 126
Cantelon, Harold, 94
car traffic, 107, 108
carbon monoxide, vii, 18, 72
Carrick, William, 45, 48
Carroll, William K., 9
Carson, Rachel, 6, 26
Casey, Sean, 85, 87
Casina, Joseph, 16, 19
Centre for Resource Recovery, 147, 151-52
Chant, Donald: and Bill Davis, 195n78; and brief to CRTC hearing, 35-36; on CELRF Board of Directors, 194n49; on diazinon levels, 46; on energy panel, 84; and Energy and Resources Project, 83; on Hall Commission, 27; participation in Survival Day, 78-79, 79; and Pollution Probe, 43, 57, 78, 101, 172, 207n2; recognition as Pollution Probe founder, 17, 207n2; vs. Solandt, 44; support for CELA, 93; support for TZPG, 70
Chant, Merle, xxiv
Chilton, Patty, 174, 175
China Syndrome, The, 138-39
chlorofluorocarbons (CFCs), xix-xx, 107-8
Citizens' Inquiry into Air Pollution, 58
Clark, John F.M., ii
Clarkson, Adrienne, 79

Clarkson, Stephen, 79
Clear Hamilton of Pollution. *See* Conserver Society of Hamilton and District
Clément, Dominique, 9, 170
climate change, xviii, xix-xx, 11
Club of Rome, 81
Collins, Donald, 79
Combined Universities Campaign for Nuclear Disarmament, 125
Community Planning Association of Canada, 114
Company of Young Canadians, 114
consciousness: ecological 3-4, 11; environmental, 37, 49, 60, 62, 64, 67, 77-78
Conservation Council of Ontario, 3
conservationism, 3, 123, 129-30; vs. environmentalism, 4-5
conserver society: ethics, 149; principles, 129-30, 133
Conserver Society of Hamilton and District, 180n2
Conserver Solution, The (Solomon), 130-32
Consumers' Gas Company, 57, 59
containers: non-returnable, 48-51, 111-12; 128, 167
Continental Can Company, 85, 87
Continental Energy Pact, 80-84
Conway, Chris, 131, 132, 145, 146
Conway, Thomas, 2
Coombs, John, 40, 42, 55
Coon, David, 142, 143
Council Organized to Protect the Environment (COPE), 92
Cox, Marilyn, 187n8. *See also* Spink, Lynn
Creed, Murray, 14, 27, 29, 30
Crooks, Harold, 147
Cumber Street Pumping Station, 44
Cunningham, James, 55
Curry, Rich, 136
Curtin, John, 155

Dales, J.H., 43
Daly, Martin, 55
Darlington Nuclear Generating Station, 136-37, 139-40, 141
David Suzuki Foundation, 169

Davidson, George F., 29
Davie, J.H., 76
Davis, Bill, 99, 100, 140, 195n78
Davis, Jack, 75, 83, 84, 90, 91
dead ducks controversy, 45-48
Degler, Teri, 174
demonstrations/protests: at Bruce Nuclear Generating Station, 136; at Darlington Nuclear Generating Station, 139-40, 141; petitions, 100, 118
Dennison, William, 47
Department of Energy, Mines and Resources (EMR), 67; Office of Energy Conservation, 132-33
Department of Energy and Resources Management (Ontario), 60
Department of the Environment, 109
Department of Health, 158
Department of Highways (Ontario), 49-50
Department of Natural Resources, 95
Department of Transport, 50
development. *See* northern development
diazinon, 39, 45-47; levels, 46
dioxin, 154-56, 158
"Do it" slogan, 51, 52, 53, 61
Doern, G. Bruce, 2
Don River: mock funeral, 1, 52-56, 53, 54, 188n48; sewage-induced bacteria levels, 52; water, 52, 53
Don't Make a Wave Committee. *See* entries starting with Greenpeace
Douglas Point, 126
Dowie, Mark, 65
Drinking Water: Make It Safe, 158
Drowley, W.B., 16
Drugge, S.E., 192n25
Dunnville, xiv-xv, 12, 15; effects of fluorine pollution, 19, 37
Dymond, Matthew, 20, 21, 22, 23, 25, 30, 31, 47

Earth Day, 6, 65, 69, 78
East York Conservation Centre, 150
Eaton, F.S., 64
Ecology Action Centre, 3, 97, 180n2, 202n52

Ecology House, 128-29, 159-60, 160, 174
Ecumenical Task Force, 155
Edmiston, Robin, viii
Edwards, Martin, 48
Edwards Inquiry, 48
Ehrlich, Paul, 69, 70, 81
Electric Reduction Company (ERCO): and CBC, 17; and CRTC hearing, 28-29; damages paid out, 16; fluoride emissions, 16, 19-20; and Hall Commission, 25-26; phosphate plant, 15, 23
energy: alternative, 123; ethics of use, 133; natural gas, 56, 57, 58, 59, 108; nuclear, critique of, 134-37
Energy Probe: anti-nuclear focus, 125-26; criticism of Porter Commission interim report, 138; and Ecology House, 128-29; and fossil-fuel industry, 145; funding, 142, 202n52; fundraising, 123-24, 143; leaders, 124-25; leaves Pollution Probe Foundation, 143-46; neoliberal agenda of, 146; and the nuclear critique, 134-37; official launch, 124; Ottawa office, 132-33; paid staff, 145-46, 146; publications, 126, 138; publicity stunts, 137-38; revenue, 126-27; sources of funding, 203n79; views of, 203n79; work against nuclear power, 126
Energy Probe action, against nuclear power, 137, 138-39
Energy Probe Research Foundation, 144; board of directors, 145; and Environment Probe, 163; expansion of, 162-63; focus on foreign development policy, 162-63; fundraising, 145; Probe International project, 163; revenues, 163; and the Urban Renaissance Institute, 163
Energy and Resources Project: agenda for Canadian government, 82; beginnings of, 75-76, 80-84; energy panel, 84, 192n25; and Energy and Resources Team, 106; public reception of, 117; and recycling, 84-91; soliciting support for, 82-83
ENGOs. *See* environmental non-governmental organizations (ENGOs)

Environment Canada guidelines, 157-58
environmental activism. *See* demonstrations/protests; environmental movement
Environmental Assessment Act: exemption granted to Ontario Hydro, 136-37
Environmental Defence Canada, 207n50
Environmental Defense Fund, 3, 93
environmental degradation: growth ethos and, 70, 76; land-use policy and, 119-20
environmental education, 3-4, 11, 32-33, 37, 49, 60, 62, 64, 67, 77-78, 104-6, 105, 120, 171, 106
Environmental Education, 106
environmental issues: global, xix, xix-xx, 36-37, 125-26, 134; local to pan-Canadian, xvi-xvii, xix; public awareness of, 171; transnational, 153, 154-57, 167; *vs.* urban issues, 116-17. *See also specific environmental issues*
environmental justice movement, 3, 7, 117, 173
environmental law, 93. *See also* Canadian Environmental Law Association (CELA); Canadian Environmental Law Research Foundation (CELRF)
Environmental Law Association. *See* Canadian Environmental Law Association (CELA)
environmental movement: Canada *vs.* US, 3-4, 6; on Canada's west coast, xiv-xv; corporate support for, 65; declining profile of, 122-23; economic recession and, xvii; and guerilla theatre, 1, 55-56; and non-violent direct action, 134; origins of, 12-13; population control and, 68-72; publications on, 2; role of summer camps in, 41, 63, 187n8; shaping, 180n4, 180n8; in Toronto, 2-3, 4; waves, 4, 8, 153, 168-71, 180n8
environmental non-governmental organizations (ENGOs): and bipartite bargaining, 172-73, 205n7; and CAHE, 91-92; Canadian, and Love Canal, 2, 204n1; competition for funding, 74; corporate support for, 65; Danish, 139; and decline of bipartite bargaining model, 11; emergence in Canada *vs.* in US, 3-4; first-wave Canadian, 5-6; of Ontario, 159; outside Toronto, 180n2; pan-Canadian, 4, 8, 75, 153, 168-69; part-time *vs.* full-time, 73-74; second wave, 167-70; stages of maturation, 101; Toronto-centric, 173; working with business, 161-62. *See also specific environmental organizations*
Environmental Protection Act, 95, 96, 118, 128, 166-67
Environmental Studies programs, 171
environmentalism: Canadian, xii; *vs.* conservationism, 4-5; Harper government and, 11; and political stances, 171; Toronto, 173
An Essay on the Principle of Population, 68
Estrin, David, 93, 95, 96-97
eutrophication, 59, 68
Everything You Wanted to Know about Nuclear Power (but Were Afraid to Find Out!), 138

Falls, J. Bruce, 36, 69
Family Herald, 26
Federation of Ontario Naturalists, 3, 48
Fiddler, Frank P., 70
Findlay, Rick, 151
Fine, Richard, 129
fluoride pollution, xiv-xv, 15, 19, 37; public inquiry into, 22, 24-27
fluorine, 19
fluorine intoxication. *See* fluorosis poisoning
fluorine pollution. *See* fluoride pollution
fluorosis poisoning, 12, 16-17, 19, 20-21, 22, 32; diagnosis of, 24, 25
Forkey, Neil S., 180n4
fossil fuels, xx, 56, 108-10; free-trade policy for, 83
fossil-fuel industry, 19, 145
free market, 131-32, 146, 162-63, 203n79; and negative income tax system, 131-32
Friends of the Earth, 1
Friends of the Earth Canada, 168, 206n48
Frye, Northrop, 28
Fuller, William, 192n25

garbage: 3Rs waste hierarchy, 111-12, 147, 167; recycling, 111-12, 163-64; reducing throughput, 50-51, 85, 86, 111, 167, 198n35
Garbage Coalition, 127
Garbage Fest 77, 163
Gathercole, George, 56, 57, 58, 59
Gaunt, Murray, 79
Gershman, Stanley, 82-83
Gibbons, Jack, 146
Gibson, Robert, 128
Gillies, Jim, 64
Gilmore, J.P., 184n36
Globe and Mail: and advertising space, 51; coverage of anti-nuclear protests, 113; coverage of CRTC hearings, 185n55; coverage of Don River mock funeral, 56; letters to the editor, 26, 49, 57, 94, 103, 161-62; on Pollution Probe anti-litter campaign, 49; Pollution Probe opinion piece, 113; and Pollution Probe public profile, 57, 65; support for Summer Project '70, 68, 95
Godfrey, Paul, 158
Golden, Alan, 28
Golden, Aubrey E., 96
Goodman, Eddie, 99, 194n49
Gosnell, Denise, xxiv, 182n5, 185n71
Gosnell, Larry: and *The Air of Death*, 6, 14, 37; awards, 15; and CRTC hearing, 27, 28, 29; and GASP, 31, 185n71; and Hall Commission, 25; *Poisons, Pests and People*, 12-13; pollution specials, 63; praise for, 185n55; at University of Toronto, xiii; and Waldbott, 16-17
Great Bear Rainforest Agreement, xxi
Great Lakes pollution, 59-60, 84, 189n70
Greed, Simon, 1, 54, 55
Green, Larry, 94, 96, 118
Green Line: controversy, 161-62, 172; products, 153
Green Party of Canada, 2
Greene, J.J., 80, 82, 83
Greenpeace Canada, 8, 168; beginnings, 100-1; defining characteristics, 10; influence of, 180n4; public profile, compared to Pollution Probe's, 177; publications on, 2; and theatrical events, 1
Greenpeace International, 168
Greenpeace Toronto, 6, 8, 134-37, 140
Greenpeace Toronto action: against Bruce Nuclear Generating Station, 136-37; against Darlington Nuclear Generating Station, 137, 140, 141
Grimsby, 89
Group Action to Stop Pollution (GASP), 5, 30-31, 31-32, 185nn71-72; actions, 39, 46-47, 72-73; brief to CRTC, 33-34; educational emphasis of, 32, 32-33; *vs.* Pollution Probe, 40
growth ethos, 70-71; and environmental degradation, 76, 81, 108

Hall, George Edward, 23
Hall, Ross, 57
Hall Commission, 23-26; report, xiii, 24, 25-26, 27, 32, 34-35
Hallett, Douglas, 155
Hallman, Eugene, 28, 29
Hamilton and Region Conservation Authority, 57
Hanes, Merle: recognition as Pollution Probe founder, 175. *See also* Chant, Merle
Harding, C. Malim, 76
Harris, W.B., 64
Hays, Samuel, 4-5, 69
hazardous waste, 8, 154-57, 157-58. *See also* Love Canal disaster
Hearn Generating Station, 56-59
Helliwell, John F., 124
Hellman, Eric, 151, 164, 165
Henderson, Gavin, 26
Henderson, Hazel, 19, 31
Higgins, Larratt, 192n25
historians, xv-xvi, xvii-xviii; treatment of environmentalism in Canada, 4-6, 9, 170-71
Hoberg, George, 11
Hodgins, J., 57
Holling, C.S., 84
Hooker Chemical Company, 154-56

Horgan, Frank, 158
Howard, Ross, 128, 140
Hudson, Clayton, 194n49
Hughes, Hubert, 46, 47-48
Hummel, Monte, 61; on diazinon issue, 48; on early days of Pollution Probe, 2, 41; and Education Team, 104; leaves Pollution Probe, 127; on Pollution Probe decision-making, 98; recognition as Pollution Probe founder, 175; and recycling, 89-90; on Tony Barrett, vi, 42; and WWFC, 168, 169-70, 207n51
Hunter, Bob, 134
Hurtig, Mel, 84

insecticides, 13-14
International Joint Commission, 59, 60
Is Five Foundation (IFF), 6, 8, 163; collaboration with DEL Equipment, 151; consulting service, 150, 151; and curbside recycling, 148, 149; employees, 150; expansion, 150-51; government funding, 149-51; as workers' cooperative, 147-49
Isaacs, Colin, 145, 153, 156, 158-60, 166, 206n31

James, Bob, 105
Jones, D.B., 13-14
Jones, Oakah, 57, 58
Jones, Phil, 43, 60
Judy, R.W., 43

Karfilis, James, 72-73
Kealey, Gregory S., 170
Keeling, Arn, 2
Kelly, Brian, 67; against Cumber Street Pumping Station, 44-45; against energy growth, 59-60, 106; and Energy and Resources project, 80-82, 82-83, 116-17, 121, 123, 132; leaves Pollution Probe, 124; and Omond Solandt, 43-44; against phosphates, 59-60, 64-65; as Pollution Probe coordinator, 62; on public appeal of *The Air of Death*, 35; recognition as Pollution Probe founder, 175; science background, 44, 62, 80; and Survival Week/Day, 78

Kelly, Ruth, 67; recognition as Pollution Probe founder, 175
Kerr, George, 50-51, 57, 58, 64
Kilbourn, William, 88
Kirkham, Bill, 165
Kitchener recycling project, 164-66
Knelman, F.H., 192n25
Kyoto Protocol, xx, 11

Laidlaw Waste Systems, 164, 165-66
Laing, Mack, 56
Laird, Arthur, 22
Lake Ontario Cement, 94-96
Lake Ontario pollution, 154-57, 157-58
Langdon, Steven, 44
Lawther, Patrick, 24, 31, 184n32
Lean, Dave, 61
"Leave the Car at Home Week," 72-73
LeDrew, Bob, 166
Lee, Dennis, 174
Lemon, James, 114
Levy, Alan, 92
Lewis, Aird, 32
Linsky, Benjamin, 58
Loblaw Groceterias: and environmental movement, 60, 68; Green Line, xvii, 153, 161-62
Local Initiatives Program (LIP) grants: cancelled, 120; to Greenpeace Toronto Education Project, 134; to Is Five Foundation, 149-50; and national unity, 91-92
Locke, Colin, 57
Love, Ann, 63, 100, 101, 111; recognition as Pollution Probe founder, 175
Love, Betty, 207n2
Love, Gage, 63, 64, 111, 207n2
Love, Peter, 63-64, 111, 112, 121, 175
Love Canal disaster, 154-57, 204n1, 205n7
Ludolph, Nyle, 163-64, 165, 166
Lytle, John Hunter, 32

MacDonald, Donald S., 70, 137
Macdonald, Doug, 2, 10-11
Macdonald, J.K., 64
MacEachen, Allan, 21, 29-30

Mackenzie Valley Pipeline, 108-10; Inquiry, 110
Maclean's, 60
Macnaughton, Alan A., 168
Macpherson, Alexander, 158
Mains, Geoff, 80-82, 83
Malthus, Thomas Robert, 68-69
Manning, Morris, 97, 195n67
Mansfield, Richard Alan, 32
Marier, Jean, 19
Marmorek, Jan, 138, 145, 146, 202n52. *See also* McQuay, Jan
Marshall, Brian, 129
Marshall, John, 149
May, Elizabeth, 169
Maynes, Clifford, 161
McAskile, Peter, 120
McCarron, Owen, xiv
McCarthy, John D., 74
McClure, Robert, 47
McCormick, John, 4-5
McDermott, Dan, 134-35
McGinnis, Jack, 147-48, 149, 151
McGough, Peter, 165
McKenna, Joseph, 58
McKinney, Alex, 23
McLaughlin, Mark J., 2
McLuhan, Marshall, 39, 43, 47
McQuay, Jan, 138, 143, 202n52
Merchant-Heeney report, 80-84
methoxychlor, 47
Metro Toronto Parks Department, 39, 45-46
Metro Toronto Public Works Department, 45
Metropolitan Toronto Airport Review Committee (MTARC), 127
Middleton, Peter, 61; background, 63; and CAHE, 91; and CELRF, 194n49; defence of Urban Team, 116; and Energy and Resources project, 83; leaves Pollution Probe, 121; against phosphates, 60; as Pollution Probe coordinator, 62, 98-99; as Pollution Probe executive director, 99-100, 104, 117-18; on Pollution Probe focus, 76, 83; and Pollution Probe legal arm, 92-93; recognition as Pollution Probe founder, 175; on revisionist history, 176; on Tony Barrett, 43; on Vickers and Benson connection, 51
Mills, Robert, 36, 41, 42, 51-52, 63, 64, 92; recognition as Pollution Probe founder, 175
Ministry of the Environment, 112, 151, 158, 199n66; Solid Waste Task Force, 197n23
Ministry of the Environment (Ontario), 96-97, 156
Moore Park project, 87-88
Morden Commission, 17, 23
Moriyama, Raymond, 64
Morrison, Barbara, 155, 157
Morse, Barry, 140
Munch, John, 140
Munro, Marcel, 25
Murphy, Tom, 115, 116
Murray, Ron, 164
Murray, Steve, 165

Naqvi, Syed, 125
National Environmental Policy Act, 48
National Film Board, 13-14
National Fluoridation News, 17
natural gas, 56, 57, 58, 59, 108
Nature Conservancy of Canada, 3
negative income tax system, 131-32
Newman, William, 113
Nixon, Doug, 14
Nixon, Robert, 61
North American Life Assurance Company, 63
northern development, 109-10
Northern Inland Waters Act, 109
Norton, Keith, 155
nuclear power: plants, 126, 136-37, 139-40, 141; public's views of, 125, 135, 136, 138, 139-40, 143; safety of, 138-39

O'Connor, Ryan, i-ii, xv-xvi, xvii-xviii, xx
O'Donohue, Tony: and GASP, 30, 31-32, 33, 46, 73, 185n72; on GASP finances, 72; and Ontario Hydro superstack, 57-58, 184n32; and Pollution Probe, 104

O'Driscoll, John, 97
O'Malley, Terry, 51, 62
Oliver, Bob, 176
Oliver, Joe, 11
Ontario Hydro, 3; account of Bruce Nuclear Generating Station security breach, 136; and air pollution, 56-59; nuclear expansion plan, 125-26, 136-37, 138; paradigm shift, 140; rate structure, 106-7; rates, 131
Ontario Ministry of the Environment, 96-97, 156
Ontario Multi-Materials Recycling Inc., 167
Ontario Paper Company, 151
Ontario Pesticides Advisory Board hearings, 47-48
Ontario Pollution Control conference, 31, 185n64
Ontario Recycling Information Service, 151
Ontario Soft Drink Association, 166-67
Ontario Waste Management Advisory Board, 112
Ontario Water Resources Commission (OWRC), 24, 67
Operation Clean – Niagara, 154-56
Opperman, Jo, 122
Opportunities for Youth, 85, 91
Organization of Arab Petroleum Exporting Countries (OAPEC), 110, 123
Osborn, Fairfield, 68
Osler, Sanford, 124-25
Ottawa, 79-80
ozone-hole issue, xix-xx, 2, 4, 8, 167

Page, Robert, 109, 197n18
Palmer, Janice, 72
Parliament Hill, 89, 90-91
Peden, William, 140
Pembina Institute, 169
Pennington, Roy, 17
People or Planes, 127
Pepper, P.B.C., 28-29
pesticides, 3, 39, 45
Pesticides Advisory Board: hearings concerning dead ducks, 47-48

petitions, 100, 118
Phillips, Mark, xviii
phosphates, x, 3, 14-15, 59-60
Pickering Airport Inquiry, 119
Pickering Nuclear Generating Station, 126
Pickett, Paul, 88
Plowright, Chris, 37, 70, 71-72, 192n25; participation in Survival Day, 78
Plowright, R.C. *See* Plowright, Chris
Poisons, Pests and People, 13-14
policy making: bipartite model, xx, 10-11, 172; multipartite bargaining and, 11
politicians: photo opportunities, 89; and Pollution Probe, 61, 61, 65; views of environmentalism, 122
pollutants: alpha-chloralose, 45, 48; benzene, 158; benzopyrene, xiii, 18; carbon monoxide, xiii, 18, 72; chlorofluorocarbons (CFCs), xix-xx; diazinon, 39, 45-47; dioxin, vii, 154-56, 158; fluorine, xiv-xv, 12, 14-17, 19, 20-22, 24, 25, 32, 37; hazardous waste, 8, 154-57, 157-58; insecticides, 13-14; methoxychlor, 47; pesticides, 3, 39, 44-45; phosphates, xvi, 3, 14-15, 59-60; sewage, 68; solid particulates, 117-18; sulphur dioxide, 56, 57, 117-18
pollution: air, xiii, xvi, 6, 18-21, 31, 34, 56-59, 189n58; control of, 21; costs of, 106-7; fluorine, xiv-xvi, 12, 14-17, 19, 20-22, 24, 25, 32, 37; from heavy industry, 115-16; measuring levels of 16, 24,72; water, xvi, 44, 52-56, 59-60, 189n70
Pollution Inquiry Committee. *See* Hall Commission
Pollution Probe: affiliates, 66, 68, 202n52; and bipartite bargaining system, 172; Board of Advisors, 43, 64, 76; board of directors, 143, 144, 206n31; centrist approach, 10; and development issues, 103-4, 114-17, 173; "Do it" slogan, 51, 52, 53, 61; energy panel, 84, 192n25; and Energy Probe, 124-25; evolution of, xvi-xvii, xxi-xxii, 2-3, 40-45, 135, 154-59, 175-76; vs. GASP and TZPG, 6-7; Greer Award, 104-5; historical memory,

175, 176; influence of, xx, 88-89, 180n4; and Loblaws endorsements, 153, 161-62; media presence, 60-68, 117, 120, 122, 126-28, 172, 177, 200n3; membership, 41, 42, 65, 207n2; and Ontario government, 99-100; organizational structure, 7-8, 40-41, 97-100, 127, 145; paid staff, 62-63, 65, 76, 87, 99, 121, 159; publications, 77-78, 80, 85-86, 103, 106, 114-15, 116, 118, 119, 147, 158, 174-75, 205n20; and recycling, 84-91, 85-87, 86, 147-52; relocation to Ecology House, 143, 144; report on Canadian Gypsum, 118; and SCA Chemical Waste Services, 157-58; seat on ACE, 107-8; and Solid Waste Task Force, 111, 112; support for *The Conserver Solution*, 130-31; support for other environmental organizations, 71, 91-92, 92-96, 97, 194n49, 195n78; and University of Toronto Department of Zoology, xx, 6-7, 34-37, 35-37, 43-44

Pollution Probe actions: against air pollution, 35, 37, 117-18; 207n4; anti-litter, 48-50, 50-51, 52-56; collaborative work, 72-73, 127-28, 154-56; for conserver society, 75-76, 78-80, 89, 90-91, 113, 115-16, 119, 127, 192n11; against Continental Energy Pact, 80-84; founding, 34-36; *vs.* Greenpeace tactics, 134; against Hooker hazardous waste, 154-56; against northern development, 109-10; against nuclear testing, 100-1; against Ontario Hydro, 57-59, 106-7, 189n58; against pesticides, 39, 45-47; against phosphate levels, 59-60; for recycling, 67-68, 75-76, 89, 90-91, 113, 128; regarding Love Canal disaster, 205n7; in Sandbanks Park case, 95-96; against unfair housing, 103-4; against water pollution, 1, 44-45, 53, 54, 157-58, 188n48. *See also specific actions*

Pollution Probe finances: budgets, 62, 76, 77, 110; business support for, 7, 43, 51-52, 53, 63-65, 68, 76-77, 199n66, 207n3; deep-rooted problems, 144-45; Ecology House and, 159-60, 174; energy crisis and, 120-21; expenditures, 62, 120, 197n22; fundraising, 142, 158-59, 176-77; government funding, 67-68, 129, 156, 199n66; Greer Award, 104-5; revenue *vs.* wages, 141-42, 159; University of Toronto and, 159

Pollution Probe Foundation: attempts to raise profile, 128-29; economic troubles of, 141-42; federal government support for, 129; fundraising, 143; opposing trajectories within, 126-29; publications, 128, 131-32, 144, 205n20; restructuring, 123; views of negative income tax system, 132

Pollution Probe Teams: Caravan Team, 120; Education Team, 104-6, 105, 171; Energy and Resources Team, 106-10, 116-17, 123, 125; Land Use Team, 119-20; Recycling/3Rs Team, 110-13; Urban Team, 103-4, 114-17, 173

population: control, 68-69, 70-71; growth, 36-37, 68-72, 70, 81

Port Maitland, 15, 16, 26, 32

Porter, Arthur, 126

Porter Commission, 126, 135, 138, 140

Power, Dennis, 71

Prairie Defence League, 202n52

Prescott, 89

Pretty, David W., 63, 76

Probe Bulletin, 116, 120

Probe International project, 163

Probe Newsletter, 77, 80, 118

Probe Post, 128, 132, 143, 144

"Probe's Last Word on 'Recycling,'" 147

Procopio, Connie, 116

Procter and Gamble, 64-65

Project One Recycling, 148-49

Pross, Paul, 101

public hearings/inquiries: into air pollution, 56-59; concerning dead ducks, 47; into Mackenzie Valley Pipeline, 109-10; into Ontario Hydro superstack, 58-59; into pesticide use, 46, 47

public interest: economic elite *vs.*, 35-36

Public Interest Coalition for Energy Planning, 202n52

Index

Purdy, David, 76

Rayner, Agda, 95, 96
Read, Jennifer, 2, 60
recycling, 84-91; collection methods, 164; and environmental deterioration, 90; experiment with, 151-52; standard program *vs.* Black-Clawson plant, 113; telephone directories, 85, 86; true cost of, 131
Recycling Council of Ontario (RCO), 151
Regier, Henry, 27, 36, 43, 69, 107, 108, 172
Reilly, Phil, 83
Resource Integration Systems (RIS), 150, 151, 164
Resources Recycling Caravan, 75-76; and Jack Davis, 89, 90, 91
Richard Ivey Foundation, 199n66
Richard L. Hearn Generating Station. *See* Hearn Generating Station
Rigg, F.D., 24
Robarts, John, 31, 60, 71, 99
Roberts, John, 155
Roberts, Kelso, 94
Robinson, John B., 130
Rocky Mountain Phosphate Plant, 15
Rothschild, Joyce, 97-98
Rounthwaite, Ann, 42, 117
Roy, L.P., 19
Royal Commission on Electric Power Planning, 126
Royal Commission on Health Services, 17
Royal Commission on Pesticides, 48
Rubin, Norm, 132, 146, 169
Rules of the Game: A Handbook for Tenants and Homeowners, 103

Sandbanks Park cases, 93-96
Sandbanks Provincial Park, 93-94
Sarnia, 19
Saskatchewan Environmental Society, 180n2
Saunders, Doug, 135, 136
Save the Environment from Atomic Pollution, 137
Save Tomorrow Oppose Pollution, 180n2

Sawma, Martin, 45
SCA Chemical Waste Services, 156
Schwass, Rodger, 15
Science Council of Canada, 129-30
Scientific Pollution and Environmental Control Society. *See* Society Promoting Environmental Conservation (SPEC)
Scrivener, Margaret, 31-32
Sedgwick, Joseph, 27
Serbent, Francis, 157
sewage, 44, 52
Sewell, John, 114, 140
Sharp, Mitchell, 83
Sheard, Joseph, 30, 32
Shields, Roy, 22
Siddall, Robert, 45
Sierra Club, 3, 65
Sierra Club Canada, 4, 8, 168-69
Sierra Legal Defence Fund, 3, 169
Siggins, Maggie, 52
Sirluck, Ernest, 47
Slocombe, D. Scott, 130
Societé pour vaincre la pollution, 180n2
Society to Overcome Pollution, 180n2
Society Promoting Environmental Conservation (SPEC), 3, 178n4, 180n2
Solandt, Omond, 17-18, 23, 26, 43-44; conflict of interests, 35
Solid Waste Management Advisory Board, 112
Solid Waste Task Force, 110-11, 112, 172
Solomon, Lawrence, 146, 203n80; and Energy Probe, 138-39, 142, 143, 144, 169; and fundraising, 145, 163, 203n79; publications, 129-32, 133, 146
Spadina Rapid Transit System, 115
Spink, Lynn, 41, 114, 115, 116, 187n8, 198n39
Spinner, Barry, 125
St-Pierre, Marc, 13
Stanfield, Robert, 56, 61
Starr, Michael, 96
Stephenson, Derek, 149, 150, 152, 164
Stewardship Ontario, 167
Stone and Webster study, 57
STOP Edmonton, 202n52
STOP Montreal, 202n52

Strong, Maurice, 130-31, 138
Stuart, Barry, 93, 194n49
sulphur dioxide, xiii, 18, 56, 57-59, 58, 117-18
Summer Project '70, 67-68, 67
Summerhayes, Frank, 103, 104
Sunderland, Philip, 156
Survival Day/Week, 78-80, 79, 192n11
Sutherland, Donald, 140

The Tail of the Elephant: A Guide to Regional Planning and Development in Southern Ontario, 119
Task Force Hydro, 106-7, 108
Taylor, James, 94, 137
Taylor, Paul, 166
Templeton, Charles, 100
Thatcher, John, 89
Therrien, Réal, 28
Thomas, C. Dana, 85, 87
Thompson, Tommy, 45-46, 47
Thomson, Roy, 42
Three Mile Island Nuclear Generating Station, 138-39
Tomlinson, Paul, 57, 62, 98; recognition as Pollution Probe founder, 175
Toronto: air pollution, 18; Borough of York, 117-18; city planning, 114; Dufferin-Davenport district, 115-16; municipality-wide paper pickup, 88-89
Toronto City Hall, 3; and "Leave the Car at Home Week," 72-73
Toronto Community Union Project, 114
Toronto Islands, 45-46, 48
Toronto Recycling Action Committee, 88-89, 172
Toronto *Star*: and advertising space, 51; on *The Air of Death*, 22; coverage of anti-diazinon campaign, 47; coverage of anti-litter campaign, 49; coverage of "anti-nuclear Woodstock," 140; coverage of Don River mock funeral, 56; letters to the editor, 26, 82-83; and Pollution Probe public profile, 65, 100, 122, 128

Toronto *Telegram*, 22, 46, 46-47, 54, 56, 57, 67, 82; and Pollution Probe advertising, 52, 53
Toronto Zero Population Growth (TZPG), 5-6, 37, 40, 70, 82; challenges of, 71-72
Toronto-Centred Region Plan, 119, 127
Total Recycling, 164, 165, 166
Tovell, Walter M., 95
Townsend, S.J., 192n25
Trainor, L., 192n25
Trudeau, Pierre Elliott, 61, 82, 189n73

United Church of Canada, 138
United States (US): and Canadian energy, 80-84; nuclear testing, 100-1; population to resource-use ratio, 81
United States Atomic Energy Commission, 100-1
University of Toronto, 69-70; activism, xv, 35, 9-10; Department of Zoology, 6-7, 35-37, 93; Pollution Patrol, 186n74; and Pollution Probe, 34-37; support for *The Air of Death*, 186n83; support for Pollution Probe, 43-44. *See also Varsity*
University of Toronto Pollution Probe. *See* Pollution Probe
Upper Humber Clean Air Committee, 117, 118
urban development: ecological/environmental issues of, 114-17
Urban Development Institute – Ontario, 103
Urquhart, Fred, 44

Van der Hoop, Robert, 45
Van Huizen, Philip, 2
Vancouver, 100
Vancouver *Sun*, 56
Varsity, viii, 34, 42, 44, 122, 186n74, 196n83, 189n58
Vaughan, J. Bryan, 64
Vickers and Benson, 51-52, 53
Vigod, Toby, 154, 155
Vogt, William, 68

Waldbott, George, 16-17, 20-21, 23, 25, 29, 183n15
Ware, Meredith, 54, 55, 56
Warecki, George, 3
Waste Diversion Ontario, 167
waste issues: chemical wastes, 156-57; complete solution, 167; hazardous waste, 154-57, 157-58; litter, 44; recycling and, 167; sewage, 48-50; study on, 85-87. *See also* garbage
water fluoridation, 17, 23
water pollution, xvi, 24, 44, 52-56, 53, 59-60, 67, 189n70
Waterloo, 89
West Coast Environmental Law Association, 169
Weyler, Rex, 134
White, Philip, 118
Whitney, Norris, 94
Whitt, J. Allen, 97-98
Whose City?, 114-15

Windsor, 89
Winegard, William C., 23
Winfield, Mark S., i, 11
Woolvett, R.H., 197n23
Wordsworth, Anne, 156, 157
World Wildlife Fund, 168
World Wildlife Fund Canada (WWFC), 4, 5, 8, 168, 169-70
Worster, Donald, 70
Wright, Janet, 143-44
WWFC. *See* World Wildlife Fund Canada (WWFC)

Yannacone, Victor, 27-28
Young, Fred, 79

Zald, Mayer N., 74
Zelko, Frank, vi, 2, 10, 195n78
Zero Energy Growth for Canada, 133
Zero Population Growth, 69, 70
Zlotkin, Stanley, 35, 62

NATURE | HISTORY | SOCIETY
GENERAL EDITOR: GRAEME WYNN

Claire Elizabeth Campbell, *Shaped by the West Wind: Nature and History in Georgian Bay*

Tina Loo, *States of Nature: Conserving Canada's Wildlife in the Twentieth Century*

Jamie Benidickson, *The Culture of Flushing: A Social and Legal History of Sewage*

William J. Turkel, *The Archive of Place: Unearthing the Pasts of the Chilcotin Plateau*

John Sandlos, *Hunters at the Margin: Native People and Wildlife Conservation in the Northwest Territories*

James Murton, *Creating a Modern Countryside: Liberalism and Land Resettlement in British Columbia*

Greg Gillespie, *Hunting for Empire: Narratives of Sport in Rupert's Land, 1840-70*

Stephen J. Pyne, *Awful Splendour: A Fire History of Canada*

Hans M. Carlson, *Home Is the Hunter, The James Bay Cree and Their Land*

Liza Piper, *The Industrial Transformation of Subarctic Canada*

Sharon Wall, *The Nurture of Nature: Childhood, Antimodernism, and Ontario Summer Camps, 1920-55*

Joy Parr, *Sensing Changes: Technologies, Environments, and the Everyday, 1953-2003*

Jamie Linton, *What Is Water? The History of a Modern Abstraction*

Dean Bavington, *Managed Annihilation: An Unnatural History of the Newfoundland Cod Collapse*

Shannon Stunden Bower, *Wet Prairie: People, Land, and Water in Agricultural Manitoba*

J. Keri Cronin, *Manufacturing National Park Nature: Photography, Ecology, and the Wilderness Industry of Jasper*

Jocelyn Thorpe, *Temagami's Tangled Wild: Race, Gender, and the Making of Canadian Nature*

Darcy Ingram, *Wildlife, Conservation, and Conflict in Quebec, 1840-1914*

Caroline Desbiens, *Power from the North: Territory, Identity, and the Culture of Hydroelectricity in Quebec*

Sean Kheraj, *Inventing Stanley Park: An Environmental History*

Justin Page, *Tracking the Great Bear: How Environmentalists Recreated British Columbia's Coastal Rainforest*

Daniel Macfarlane, *Negotiating a River: Canada, the US, and the Creation of the St. Lawrence Seaway*

John Thistle, *Resettling the Range: Animals, Ecologies, and Human Communities in British Columbia*

Printed and bound in Canada by Friesens
Set in Garamond by Artegraphica Design Co. Ltd.
Copy editor: Frank Chow
Proofreader and indexer: Dianne Tiefensee